Federal Data Science

Federal Data Science

Transforming Government and Agricultural Policy Using Artificial Intelligence

Edited by

Feras A. Batarseh

Ruixin Yang

Academic Press is an imprint of Elsevier
125 London Wall, London EC2Y 5AS, United Kingdom
525 B Street, Suite 1800, San Diego, CA 92101-4495, United States
50 Hampshire Street, 5th Floor, Cambridge, MA 02139, United States
The Boulevard, Langford Lane, Kidlington, Oxford OX5 1GB, United Kingdom

Notices
Knowledge and best practice in this field are constantly changing. As new research and experience broaden our understanding, changes in research methods, professional practices, or medical treatment may become necessary.

Practitioners and researchers must always rely on their own experience and knowledge in evaluating and using any information, methods, compounds, or experiments described herein. In using such information or methods they should be mindful of their own safety and the safety of others, including parties for whom they have a professional responsibility.

To the fullest extent of the law, neither the Publisher nor the authors, contributors, or editors, assume any liability for any injury and/or damage to persons or property as a matter of products liability, negligence or otherwise, or from any use or operation of any methods, products, instructions, or ideas contained in the material herein.

Library of Congress Cataloging-in-Publication Data
A catalog record for this book is available from the Library of Congress

British Library Cataloguing-in-Publication Data
A catalogue record for this book is available from the British Library

ISBN: 978-0-12-812443-7

For information on all Academic Press publications visit our website at
https://www.elsevier.com/books-and-journals

Publisher: Andre G. Wolff
Acquisition Editor: Nancy Maragioglio
Editorial Project Manager: Barbara Makinster
Production Project Manager: Punithavathy Govindaradjane
Designer: Christian J. Bilbow

Typeset by TNQ Books and Journals

To our families, as they continue to inspire us, but will probably not enjoy reading this book!

Also to you, the curious reader; may you believe that through better data, there can be a better tomorrow—whatever that means.

The editors (Feras and Ruixin)

Contents

3. Machine Learning for the Government: Challenges and Statistical Difficulties

Samuel Eisenberg

4. Making the Case for Artificial Intelligence at Government: Guidelines to Transforming Federal Software Systems

Feras A. Batarseh and Ruixin Yang

Section 2
Governmental Data Science Solutions Around the World

5. Agricultural Data Analytics for Environmental Monitoring in Canada

Ted Huffman, Morten Olesen, Melodie Green, Don Leckie, Jiangui Liu and Jiali Shang

List of Contributors

Feras A. Batarseh George Mason University, Fairfax, VA, United States

Zhongxin Chen Institute of Agricultural Resources and Regional Planning, Chinese Academy of Agricultural Sciences, Beijing, China

Tammy Crane U.S. Department of Defense, Norfolk, VA, United States

Manish Dashora George Mason University, Fairfax, VA, United States

Samuel Eisenberg Jersey City, NJ, United States

Candace Eshelman-Haynes NATO Allied Command for Transformation, Norfolk, VA, United States

Jay Gendron Booz Allen Hamilton, Washington, DC, United States

Melodie Green Agriculture and Agri-Food Canada, Ottawa, ON, Canada

Jose M. Guerrero Infoseg, S.A., Barcelona, Spain

Ted Huffman Agriculture and Agri-Food Canada, Ottawa, ON, Canada

Zhiwei Jiang National Meteorological Information Center of China, Beijing, China

Erik W. Kuiler George Mason University, Arlington, VA, United States

Henri Laude Laboratoire BlueDxX, BlueSoft Group and Advanced Research Partners, Charenton le Pont, Paris, France

Don Leckie Natural Resources Canada, Victoria, BC, Canada

Changan Liu Institute of Agricultural Resources and Regional Planning, Chinese Academy of Agricultural Sciences, Beijing, China

Jiangui Liu Agriculture and Agri-Food Canada, Ottawa, ON, Canada

Connie L. McNeely George Mason University, Arlington, VA, United States

Steve Mortimer Dominion Enterprises, Norfolk, VA, United States

Morten Olesen Agriculture and Agri-Food Canada, Ottawa, ON, Canada

Haizhu Pan Institute of Agricultural Resources and Regional Planning, Chinese Academy of Agricultural Sciences, Beijing, China

Gowtham Ramamoorthy George Mason University, Fairfax, VA, United States

Mirvat Sewadeh US Department of Agriculture, Washington, DC, United States

Jiali Shang Agriculture and Agri-Food Canada, Ottawa, ON, Canada

Jeffrey Sisson US Census Bureau, Washington, DC, United States

Min Sun George Mason University, Fairfax, VA, United States

Ruixin Yang George Mason University, Fairfax, VA, United States

Manzhu Yu George Mason University, Fairfax, VA, United States

About the Editors

Feras A. Batarseh is a Research Assistant Professor in the College of Science at George Mason University (GMU). His research spans the areas of Data Science, Artificial Intelligence, Federal Data Analytics, and Context-Aware Software Systems. Dr. Batarseh obtained his PhD and MSc in Computer Engineering from the University of Central Florida (UCF) (in 2007 and 2011) and a Graduate Certificate in Project Leadership from Cornell University (in 2016). His research work has been published in prestigious journals and at international conferences, as well as in several book chapters. He taught data science and software engineering courses at multiple universities, including GMU, UCF, and George Washington University. Before joining GMU, Dr. Batarseh was a Program Manager with the Data Mining and Advanced Analytics team at MicroStrategy. Throughout his career, Dr. Batarseh worked with multiple federal and state government agencies on a variety of data science applications and analytical deployments. In addition, he helped several organizations develop intelligent systems, make sense of their data, and gain insights into improving their operations. He is currently working with the Economic Research Services (at the US Department of Agriculture) toward building intelligent data management systems. He is a member of IEEE, ACM, BICA, and STP professional communities. For more information and contact details, please refer to his Web page: https://cos.gmu.edu/ggs/people/faculty-staff/feras-batarseh-2/.

Ruixin Yang is an Associate Professor in the Department of Geography and GeoInformation Sciences – College of Science at George Mason University. He received his PhD in Aerospace Engineering from University of Southern California in 1990. His research areas ranged from Fluid Dynamics to Astrophysics and General Relativity to Data Science, Data Information Systems, Data Analysis, Data Mining, and Earth Systems Science. On software engineering, Dr. Yang led a software development team that built several prototypes for earth science data information systems. The systems include Virtual Domain Application Data Center (VDADC) prototype system for online Earth science data analysis through Web technology; Seasonal to Inter-annual Earth Science Information Partner (SIESIP) online data search and analysis system, which integrates Oracle Database Management System, Java web technology, and GrADS, a data analysis and visualization software package; and an XML-based Distributed Metadata Server (DIMES) for representing, storing, retrieving, and interoperating metadata in a distributed environment. Dr. Yang's research is focused on exploratory data analysis and/or data mining with advanced data analysis methods to hurricane-related earth science data. He has published several referred papers on earth science data search, access, online analysis, metadata management, content-based search, bid data analytics, and data mining for rapidly intensifying tropical cyclones. He can be reached at: ryang@gmu.edu.

Note From the Editors

If you are a federal analyst, a federal engineer on a data science team, an industry consultant building big data analytics for the government, a student learning about policy making or artificial intelligence, an agricultural economist, a data scientist, an interested researcher, or if you are involved in any form of science and technology policy making, this book is for you.

A group of experts, professors, researchers, and practitioners gathered to write this book. They came from a variety of affiliations, such as George Mason University, George Washington University, US Department of Defense, The Census Bureau (US Department of Commerce), Economic Research Service (US Department of Agriculture), NATO Allied Command for Transformation, National Science Foundation's Spatiotemporal Innovation Center, Agriculture and Agri-Food Canada, Canadian Forest Service (Natural Resources Canada), Chinese Academy of Agricultural Sciences (Ministry of Agriculture), National Meteorological Information Center of China, Booz Allen Hamilton, BlueSoft Group & Advanced Research Partners (France), Infoseg (Spain), and Dominion Enterprises.

Multiple books have been published in the areas of data analytics, government software, intelligent systems, and policy making. This book, however, is at the very nexus of these topics. We invite you to explore it, and join us in our efforts to advance policy making at the government through data science.

Feras A. Batarseh and Ruixin Yang
George Mason University
Fairfax, Virginia
2017

Foreword

Data can appear lifeless and dull on the surface—especially government data—but the thought of it should actually get you excited. First off, data are exactly the stuff we bother to write down—and for good reason. But their potential far transcends functions such as tracking and bookkeeping: data encode great quantities of experience, and computers can learn from that experience to make everything work better.

For example, take agriculture—and the federal studies that advance it. A farmer jots down crop rotations as needed to manage operations. And later these data also serve to track crop productivity. But for the real payoff, the farm boosts productivity rather than only track it. By number-crunching its records, it discovers ways to optimize operations: which crop schedule, seeds, treatments, irrigation options, fertilizers, and other process decisions best increase crop yield. By learning from the data, the farmer notes that notes are more notable than previously noted.

The need for optimization is palpable. By the end of the century, agriculture will have to deal with a population of 11 digits—a vast number of mouths to feed—in combination with a diminishing supply of available land. This era of exploding magnitudes decidedly demands data science. Yet the growth in scale is actually good news for predictive analytics. It presents a greater opportunity than ever, for two reasons. First, it means more "fuel for intelligence," i.e., more data from which to learn. And second, larger-scale operations themselves stand to benefit that much more when optimized—the returns are commensurate.

This principle applies far beyond agriculture, manufacturing, and even the industrial sector as a whole. Data science drives growth and efficiency across verticals, including financial services, insurance, retail, e-commerce, energy, and health care, bolstering business functions such as sales, marketing, advertising, customer service, human resources, risk management, and supply chain management.

In government, data science's vital impact extends just as far and wide, empowering agencies to more effectively serve and safeguard citizen fundamentals, such as health, safety, housing, economic stability, education, equality, and justice. Here are some more specific areas where predictive analytics bolsters the effectiveness of government.

Health and safety: Government agencies target which buildings, restaurants, and manholes to inspect by predictively modeling which have the greatest risk of fire, health code violations, lead paint poisoning, or other safety incidents. The Environmental Protection Agency plans to use predictive analytics to regulate air emissions, monitor water quality, and avert environmental catastrophes. And the Centers for Disease Control and Prevention applies predictive modeling to improve population health.

Law enforcement: As is well known, police analytically predict crime, and judges and parole boards pay heed to recidivism risk scores. But a range of other agencies also employ data science to enforce laws and regulations. New York City analytically flags for possible illegal apartments, misused business licenses, and other ducked regulations. And the pertinent departments predictively target fraud auditing of tax returns, government invoices, government contracts, workers' compensation, and Medicaid and Medicare claims. Florida's Department of Juvenile Justice determines rehabilitation assignments based on the predictions of future repeat offenses. And other analytical efforts target internal investigations of potential police misconduct and other forms of injustice.

Defense and homeland security: Military agencies analytically predict threats and civil unrest, whereas the National Security Agency and Federal Bureau of Investigation predict terrorism. Detecting possible hacker or virus footprints toughens cyber security. The US Department of Defense applies data science to target critical internal operations such as Special Forces recruitment (predictive hiring decisions), proactive veteran suicide intervention, and the maintenance of those Army vehicles at a higher risk of impending failure.

Predictive analytics is the Information Age's latest evolutionary step. We have moved beyond engineering infrastructure that stores and manages big data to implementing science that makes actionable use of the data, tapping its contents to optimize most every large-scale activity. The breadth of examples listed previously signals that predictive analytics' role is well established, a status that is further upheld by the many other application areas we see covered at the Predictive Analytics World for Government conference (an offshoot of the PAW event series for which I serve as founder).

But the fortification of government with data science has only just begun. The pressure mounts as citizen needs intensify, international competition escalates, and infrastructure and security risks grow. Critical measures for alleviating these pressures include reducing waste and abuse, increasing the effectiveness of both targeting and triaging, and optimizing operations for efficiency. Data driven optimization is a key method for achieving these improvements.

To more fully broaden the role of data science in government—and thereby seize the tremendous opportunity of today's data eruption—agencies must collaborate. Given the complexity of both managing and analyzing big data, the "use your data!" rallying cry must not only mobilize analytics internally but also call for sharing technological resources and best practices, for coordinating efforts, and for investing in data interoperability. If a small farm has limited data, it relies on cooperation that pulls together data across many farms.

To aptly serve the needs of citizens, government agencies must advance and expand the deployment of data science. If you will allow a mixing of metaphors, you can bet the farm on tools that harvest insights from data and cultivate prosperity. And as you take your next steps in contributing to this historic development, this book guides the way—it has been ideally crafted for that very purpose by a select, international group of experts who come from a diverse range of government and industry backgrounds.

—Eric Siegel, PhD, founder of the Predictive Analytics World conference series, which includes PAW Government, and author of *Predictive Analytics: The Power to Predict Who Will Click, Buy, Lie, or Die* (Revised and Updated Edition).

Preface

In 1822, the founding father James Madison said: "A popular government, without popular information, or the means of acquiring it, is but a prologue to a farce or a tragedy, or perhaps both." Recent technological waves have evidently served Madison's vision of government transparency. The latest advancements in Artificial Intelligence (AI), Data Science, and Machine Learning can make federal data openness a low hanging fruit. Moreover, the big data and open government initiatives (signed in 2012 and 2013 by President Obama) are major enablers for transforming government into a new era of intelligent and data-driven policy making.

Madison advocated for government information openness, whereas Thomas Jefferson emphasized the importance of limited government as vital to the advancement of sciences. That Jeffersonian notion was renewed in 1961 by President Eisenhower. In his farewell speech, he warned against the domination of science through high federal involvement and big government funding. The "military-industrial complex" (a term coined by Eisenhower himself) is the perfect example of the domination that drives scientific research in the United States. Consequently, similar to how the wall of separation between Church and State was accentuated, Jefferson would have introduced a principle of that sort to keep Science fairly independent from the hands of the State.

The evolution of research, from a pursuit of educated men with curiosity to a multimillion dollar business, provides the need for some form of "management of science." Therefore, some argue that government involvement is inevitable, especially that federal funding has led to multiple success stories in science, such as the Space Program (supported by President Kennedy), Manhattan Project, Navy Transit, Solar Decathlon, and many others. Many countries have ministries of science (that fund research), China, India, Brazil, and Canada to name a few. They all have a dedicated minister of science as well. In the United States, however, multiple government agencies fund research, including the National Institute of Health, the Defense Advanced Research Projects Agency, the National Science Foundation, and others. Such agencies provide large amounts of funding to researchers, universities, and national laboratories across the country. Furthermore, as far back as the birth of this nation, when George Washington was interested in advancing the field of agriculture, many universities were supported (or founded) through federal land grants to study different areas of agriculture (such as Cornell and Purdue). Other universities were funded to advance the research of oceans and seas, in what was called a sea grant (George Mason University and the University of Delaware). Using the same model, space grant universities were funded as well (i.e., University of Hartford and the University of Central Florida). Most recently, sun grants were created for the advancement of renewable energy research (i.e., Pennsylvania State University). Hence, and as it is evident, government spending leads to impelling what areas of research thrive more than others.

The public, however, is not always clear on what fields are being funded and why. In a free market system, free enterprise can drive the demand for certain sciences, but when the government is the main funder of science, researchers will pursue what the government is looking to fund. Society could be impoverished by this process, especially that some do not trust the government because of bureaucracy, wasteful spending, transparency, and accountability issues. More importantly (and as this book establishes), problems exacerbate when government is contracting with the private sector for the development of software projects. For example, the Departments of Energy (DoE), Defense (DoD), and Agriculture (USDA) spend ample amounts of tax payers' money on software research and development contracts. A reductionist approach to dealing with this issue would advocate for that government should be completely out of contracting and scientific research. Yet, it is more realistic to provide holistic solutions to this very important debate (therefore, this book), such as increasing data openness, lobbying government to being more transparent with the public through sharing raw data sets, and creating scientific methods (not driven by politics or profits) that aid in policy making. This book aims to illustrate that when government utilizes AI and data openness, it would reflect higher accountability, better governance, advances in citizen services, reduced wasteful spending, improved policy making, and the overall good of the keen American public (and possibly the world). This work is a trip through multiple AI methods, guidelines, data science use cases, and examples that are focused on federal processes, policies, and software systems.

As the size of data sets surge, the need for quicker processing, improved algorithms, better visualizations, and increased intelligence becomes obligatory. Authors of this book describe the current state of federal data science and discuss the main challenges, risks, and difficulties of injecting AI into the government. In addition, they introduce similar experiences from countries around the world, scientific solutions, and future directions.

It is also important to note that the fields of data science and AI are not all sunshine and roses. Highlighting the benefits of AI to humans should reduce the fears that many researchers and thought leaders (such as Stephen Hawkins and Elon Musk) focus on. The arguments against AI have been mostly driven from the fears of machine superiority, and the inherited complexity of the human brain that is characterized in areas such as psychology, philosophy, and biology. From the early days of AI research, many argued against the possibility of a complete and general machine intelligence. One of the strongest arguments is that it is not yet clear how AI would be able to replicate the human brain and its biochemistry; therefore, it will be very difficult to represent feelings, thoughts, intuitions, and moods in a machine.

The question posed by Alan Turing (the father of AI) still remains unanswered "Can Machines think?" It requires the creation of multiple subquestions; many fields of research could contribute to part of the answer. For example, the field of philosophy needs to study the following question: are humans intelligent enough to create intelligence? The fields of math and psychology could answer questions similar to this: could numbers and mathematical patterns represent human cognition and emotions? Also, computer hardware experts can consider this question: how much hardware power is needed to create a generic AI machine? AI therefore, is not a slim challenge; rather, it is a wide and complex field of study. Consequently, for AI and data science to reach their intended goals, they must be studied from both the technical and nontechnical perspectives. One of the goals of this book is to make the case that by injecting AI into government software (starting with agricultural systems) AI and data science can have a great deal of positive effect on our lives.

The work presented in this book hopes to ignite the engines of innovation at many information technology divisions of the government and create a major dent in the argument that suggests AI is a dangerous pursuit. However, to create that reality, a wide array of challenges needs to be investigated (such as users' adoption, trust in the outcomes and skill levels), ideas need to be researched, tools need to be developed, and scientific solutions need to be presented. That is precisely the premise and promise of this book: to provide the academic, technological, and technical means to empower the government to transforming from a government of bureaucracies and legacy software systems to a government of data-driven processes and outstanding intelligence; a difficult pursuit, but a compulsory one.

Feras A. Batarseh
Research Assistant Professor, College of Science, George Mason University
Adjunct Professor, School of Engineering and Applied Science, George Washington University
Washington, DC
2017

Section 1

Injecting Artificial Intelligence Into Governmental Systems

Chapter 1

A Day in the Life of a Federal Analyst and a Federal Contractor

Feras A. Batarseh
George Mason University, Fairfax, VA, United States

In a country well governed, poverty is something to be ashamed of; in a country ill governed, riches and honors are to be ashamed of.

Confucius

1. IN THE EARLY MORNING

Every morning, as some federal employees shuffle through the lanes of Independence and Constitution Avenues, and as some others walk out of the orange, blue, red, green, or yellow line metros, they are thinking about today's task: how to answer the news reporter who sent them a request this morning, or how to respond to the request by a senator, how to update the agency's website to handle the public's demands, or even, in some cases, how to address an inquiry by the Pentagon or the White House. Such requests can require a variety of skills and knowledge, but one thing that they all share is the need for valid, dependable, clear, complete, and insightful *data*.

Government spends mammoth amounts of money (on federal projects and federal contractors) to provide data to federal analysts (Yu, 2012; NIST, 2012). This chapter aims to assess that process, evaluate government spending with software and data engineering projects, and underline the major players in federal contracting and the main federal agencies that are expending on such endeavors.

Often, the public looks through different governmental websites to obtain the required data (Bertota et al., 2014). Such data are generated from within federal departments, shared among different agencies, analyzed by federal analysts, and *partly* published to agencies' websites. Fig. 1.1 shows the number of monthly website visits for major departments (Joseph and Johnson, 2013). The departments of Education, Agriculture (which is heavily addressed in this book), and Veteran Affairs witness the most action. That seems to be consistent with

Federal Data Science. http://dx.doi.org/10.1016/B978-0-12-812443-7.00001-6

Department	Website	Employees (2010)	Sources of big data	
			Monthly site visitors* (Quantcast.com)	Sites linking in* (Alexa.com)
1. Agriculture	usda.gov	98,235	2,800,000	69,658
2. Commerce	commerce.gov	45,348	7,700	42,281
3. Defense	defense.gov	771,614	441,100	10,237
4. Education	ed.gov	4,611	2,900,000	53,467
5. Energy	energy.gov	16,651	501,400	24,334
6. Health and Human Services	hhs.gov	83,745	173,200	35,950
7. Housing and Urban Development	hud.gov	9,818	789,700	32,376
8. Homeland Security	dhs.gov	191,197	680,600	21,280
9. Interior	doi.gov	72,168	41,100	6,550
10. Justice	justice.gov	118,104	446,000	19,488
11. Labor	dol.gov	16,554	547,400	23,760
12. State	state.gov	12,086	1,800,000	61,586
13. Transportation	dot.gov	58,189	841,300	36,432
14. Treasury	treasury.gov	112,541	308,100	8,053
15. Veteran Affairs	va.gov	312,878	2,700,000	33,798

**Data collected in April 2013*

FIGURE 1.1 Number of website visits (Joseph and Johnson, 2013).

the number of external sites linking in to federal websites as well. Organizing data for these websites can be a daunting task.

The open and big data initiatives signed by the Obama administration have pushed agencies and departments to share their data publicly. That resulted in having data available on websites such as data.gov, the Census Bureau, and the PACER system (public courts online access system). BLS (Bureau of Labor Statistics) and NASS (National Agricultural Statistics Service) are major agricultural sciences examples of federal data system that adopted the open data initiative and made their data public to citizens (US Department of Agriculture Website's Data Products). States and cities started publishing data to the public as well. For example, the cities of Chicago (City of Chicago Website's Data Products) and San Francisco (City of San Francisco Website's Data Products) both have open data portals, *data.sfgov.org* and *data.cityofchicago.org*, respectively. That, however, led to an instant increase in the need for better database systems, big data technologies, and intelligent software systems. The government has long latched onto its *legacy systems*, and still spends ample amounts of time, money, and manpower on their maintenance. Many would argue that if the government is incrementally moving toward state-of-the-art software and data systems, then it is just a matter of time before the government catches up with the industry and it will eventually be on the cutting edge of technology. Well, to evaluate that, let us take a look at what the government is doing in terms of software and data systems; the remaining of this chapter observes different federal departments, different performance metrics, and the current state of data science, big data, and Artificial Intelligence (AI) at the US government.

2. LATER IN THE AFTERNOON

After grabbing lunch from one of Washington D.C.'s weekday food trucks, the federal analyst returns to his office and starts getting ready to meet with the federal contractor who was hired last year to test, optimize, improve, build, or maintain the agency's software system or data warehouse. Contractors differ in how they perform engineering tasks at the government. Data have been collected from an assortment of federal resources (IT Dashboard Governmental Data Portal, 2016) to evaluate contractor engagements across the US government. The data sets included information on all the major software government projects in the last 6 years, all the contractors, contract ID numbers, contracts' monetary values, contracts per agencies, time information, performance measures, investment information, vendor names, time consumed, and costs (IT Dashboard Governmental Data Portal, 2016). For all the raw or cleaned data and the data sets used in this chapter, please contact the author (also available here: *itdashboard.gov*).

The first data set showed that the Department of Defense (DoD) has the highest number of contracts, which is not a surprise whatsoever. Table 1.1 shows a comparison of defense versus nondefense spending on contractors for software systems, big data, and information technology. In a 2016 study performed by McKinsey & Company, it was surprising to show that the Pentagon hid wasteful spending in the amount of $125 billion dollars (on software projects and data), a number that was also published by the Washington Post (Whitlock and Woodward, 2016).

Fig. 1.2 shows the total *number* of major contracts per department from 2010 to 2016, and again, the DoD is the highest by far (followed by Treasury and Homeland Security). However, even if it is assumed that these departments are the most critical ones, and are the ones on the bleeding edge of innovation and development, quality control is still a major concern. How is performance on these contracts evaluated? Or in other words, how is government showcasing accountability?

Based on data from federal websites, there are two types of contracts when it comes to measuring quality, ones that are based on performance and delivery (called performance-based contracts), and others that are just not (non-performance-based contracts). Fig. 1.3 shows the total number of performance-based contracts per department.

TABLE 1.1 DoD versus Non-DoD Spending on Data Science and Software

Spending/Year	2015	2016	2017
DoD spending	$36,727 million	$37,987 million	$38,551 million
Non-DoD spending	$49,965 million	$50,726 million	$51,300 million

DoD, Department of Defense.

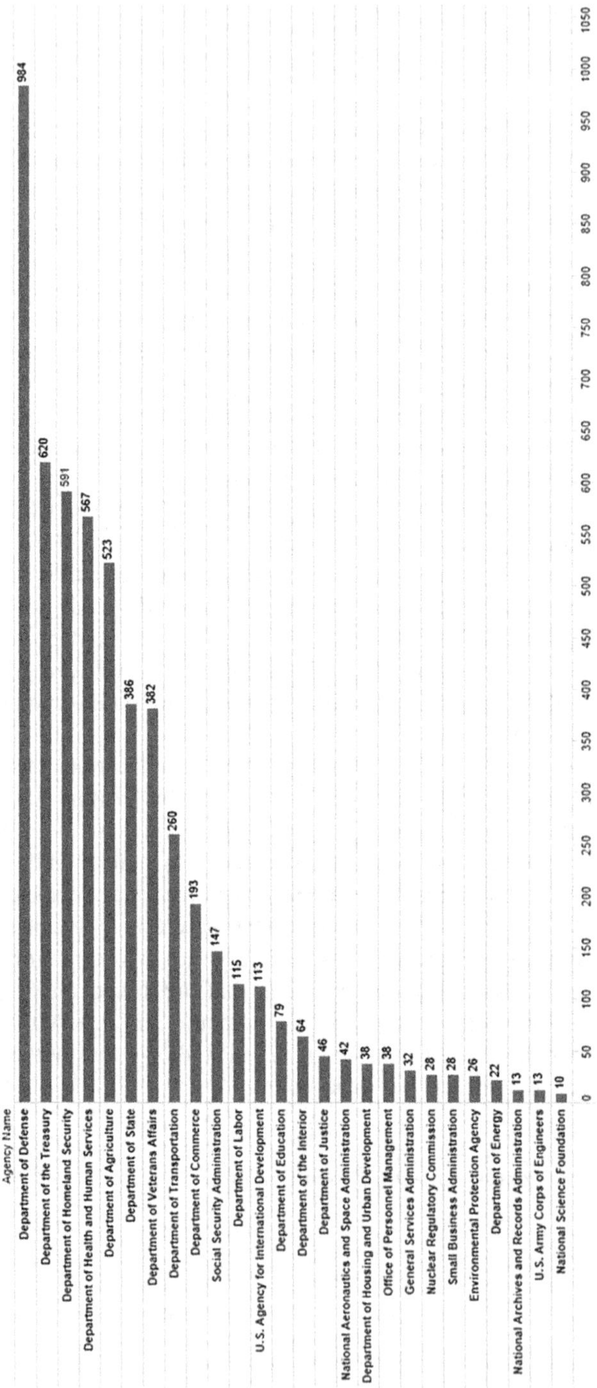

Agency Name	
Department of Defense	984
Department of the Treasury	620
Department of Homeland Security	591
Department of Health and Human Services	567
Department of Agriculture	523
Department of State	386
Department of Veterans Affairs	382
Department of Transportation	260
Department of Commerce	193
Social Security Administration	147
Department of Labor	115
U.S. Agency for International Development	113
Department of Education	79
Department of the Interior	64
Department of Justice	46
National Aeronautics and Space Administration	42
Department of Housing and Urban Development	38
Office of Personnel Management	38
General Services Administration	32
Nuclear Regulatory Commission	28
Small Business Administration	28
Environmental Protection Agency	26
Department of Energy	22
National Archives and Records Administration	13
U.S. Army Corps of Engineers	13
National Science Foundation	10

0 50 100 150 200 250 300 350 400 450 500 550 600 650 700 750 800 850 900 950 1000 1050

FIGURE 1.2 Major contracts per federal department. *Data from itdashboard.gov.*

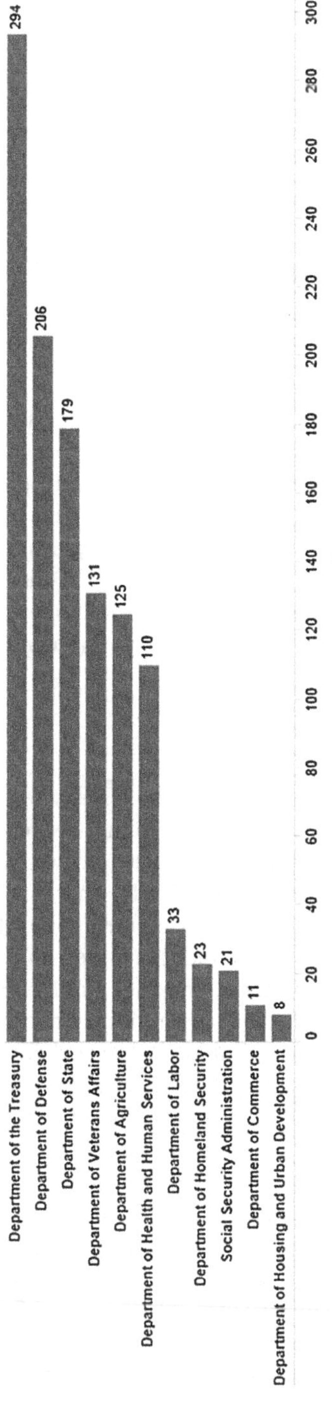

FIGURE 1.3 Performance-based contracts per federal department.

Unless it is not reported sufficiently, and as it is evident from Fig. 1.3, very few departments apply "reported" performance metrics on contracts. Based on the data, of the total **4912** major federal software engineering contracts from 2010 to 2016, only **448** contracts were performance-based, the rest are non-performance-based. Is that a bad sign? Well, given that quality could be a qualitative issue (IBM, 2007), let us consider a quantitative measure: the costs of contracts in relation to performance and quality:

1. Performance-based contracts accrued a sum of $2564.32 million in spending with an average of **7.6 million** per project.
2. Non-performance-based projects accumulated a *much higher number*; such projects have an average of **49.8 million** in spending (IT Dashboard Governmental Data Portal, 2016).

Given this big difference in spending, a major question is raised, why is not performance always considered? Where is the information (data) that shows what goals have been accomplished by these costly projects? No data were found for that. It is merely observed that to create a change at a federal agency, *the culture needs to change*, as well as the people in some cases. The next section digs deeper into the winners and losers of this seemingly inefficient contracting process. The remaining of the book proposes solutions and directions toward increasing data openness and government transparency through data science and AI.

3. LATE, LATE AT NIGHT

By 5.00 p.m., most federal employees have already left work, most software backups, data security protocols, and operating systems updates occur overnight. Federal employees use all types of software tools and systems, and many software vendors (such as Microsoft, Oracle, and Hewlett Packard) provide frequent updates to software that the government uses. How do these systems get deployed and who deploys them though? Based on the data collected from federal sources, only **1100** entities (companies and laboratories) constitute the contractors that the government deals with for major software and data science projects. Since 2010, the government has contracted **5360** major projects with the 1100 private sector contractors. Digging deeper, however, **5354** projects are performed by the private sector, and only **6** major projects are found to be designated to American universities (*begs the question, why?*). Contracting with universities and research laboratories has many advantages: universities charge much less, employs scientists and developers who are not driven by their bonuses or high salaries, and help the federal agencies avoid depending on one set of tools or software vendors in its deployments—universities are vendor agnostic.

Fig. 1.4 is a bubble chart visualization developed based on the data from federal sources. The illustration shows the relationship between federal departments and the 1100 contractors. By observing the bubbles in the figure (showing data from 2010 to 2016), although it has been already established

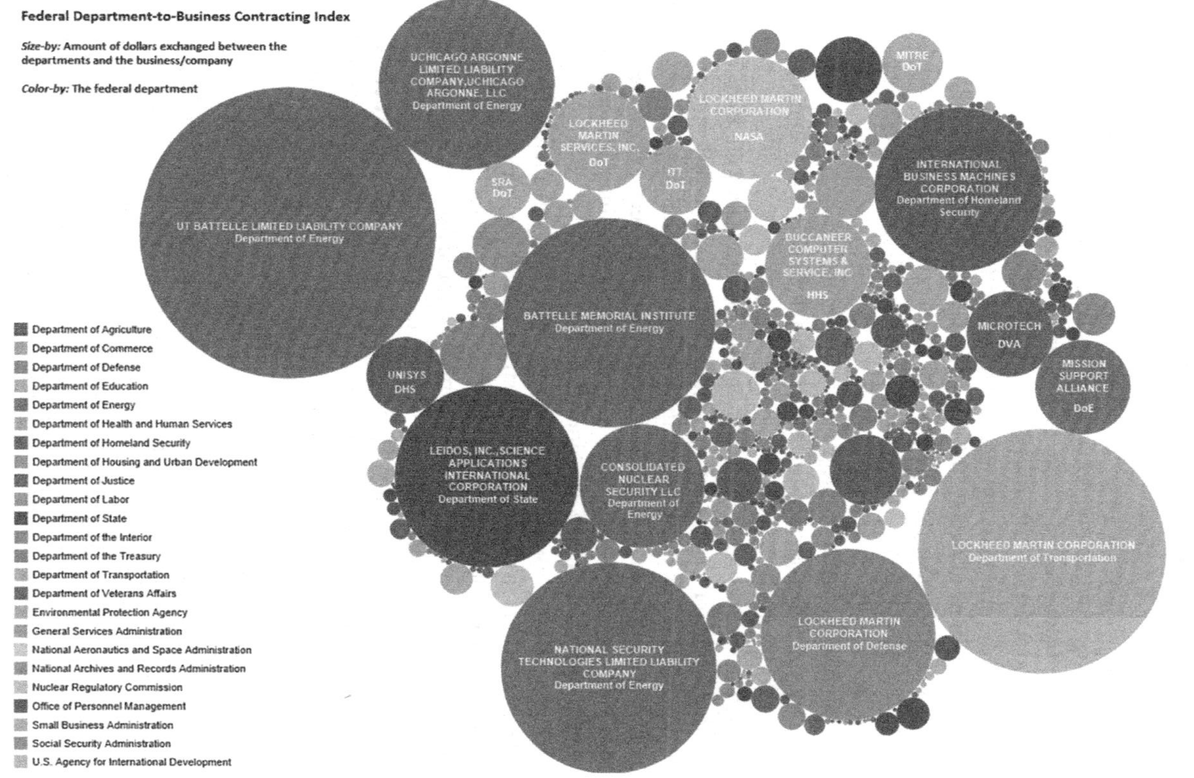

FIGURE 1.4 Federal departments to contractors Index.

that DoD is the biggest contractor, it is clear that the Department of Energy, shown by the green bubbles, has the largest deals with contractors (the chart is colored by department and sized by the cost of the project). Other departments such as Homeland Security and Transportation also have very large contracts, larger than most other departments. Other departments tend to spread out the funds on more contractors (which could induce competition and be a better practice). The contractors, when they apply certain solutions, make the agency highly dependent on them, as they inject certain tools that only they can maintain or update. Furthermore, contractors come to the agency's work site, build, deploy, and take off, which leaves the federal engineers in most cases clueless on how to manage or maintain the new systems without extensive training (that is also managed by the same contractor in most cases). Therefore, the agency becomes imprisoned to the wills, timetables, costs, and design decisions of the contractor.

More information on these projects' success/failure (measured by time and budget metrics), broken by department, is shown in Table 1.2. For complete raw data sets, please contact the author.

Although it is apparent that a finite number of data sets were used in this study, and although other data sets or studies might point to different percentages and outcomes, it is obvious that there is a dire lack in data that can point to the outcome of software and data-related federal projects. How they were evaluated and how much money will be spent on their maintenance and future development are among many questions to be asked.

Although some success percentages are presented on some federal websites for the general public to observe, there is still a lack of raw, clean, and normalized data (that can enable the curious minds to analyze more). Most data sheets provided have many empty data cells, but most importantly, most of the data are presented in formats such as PDF and MS Word, which are not useful for data analysis. Furthermore, the quality of the data presented on federal websites is usually low, and in most cases aggregated to high levels, which also makes it difficult to flexibly extract insights from.

4. THEREFORE, THIS BOOK

Although most of the contractors who work for the government have special teams working on federal projects, the mentality and science of engineering at the private sector is driven by a different set of attributes than the government, such as (*private sector vs. federal government*): different goals (*corporate profit vs. domestic tranquility*), different missions (*corporate competition vs. citizen services*), different decision actors (*few at a company vs. many at government*), and different financial resources (*revenue vs. taxes*). The question that this discussion leads to: is it fitting that most contracts are signed with the private sector? What about universities, national laboratories, and other scientific and educational institutions around the country? Such a shift towards other institutions is encouraged.

TABLE 1.2 List of Projects Success Percentages per Federal Department

Department/ Spending on Major Projects	On Schedule (%)	On Budget (%)	Total of Major Investments
USDA	61	81	32
Commerce	68	61	29
DoD	61	81	121
HHS	83	77	108
Interior	81	68	39
Justice	81	82	20
Labor	80	83	19
State	77	77	19
Treasury	82	62	59
Education	76	81	40
DoE	91	80	12
Transportation	75	73	32
DHS	79	81	92
Housing and Urban Development	73	67	5
Veteran Affairs	74	44	24

DHS, Department of Homeland Security; *DoD*, Department of Defense; *DoE*, Department of Energy; *HHS*, Health and Human Services; *USDA*, US Department of Agriculture.

The authors of this book present academic and research-driven (vs. profit driven) solutions, methods, and guidelines that would greatly improve data management processes at the government, reduce spending, increase software efficiency, validate and verify federal systems, and eventually make the day of the federal analyst much more data driven, productive, and beneficial to the goals of government.

The major fuel for AI algorithms, besides the algorithm itself, is data. These days, however, where are these data coming from? Basically, any button we click, any switch we flip, anything we sell, buy, or try, is a data point that can drive profit and decision making. Regarding government, many pointed to federal programs that collect and use our data, such as: Vault 7 and PRISM (Wikileaks Reports, 2017). However, can research centers in the United States, faithful citizens, or curious analysts access *public* federal data electronically? Currently, the answer is no. This book points to many challenges on why that is not currently possible. Federal agencies need to reevaluate their data science practices to make their raw data (not aggregated or filtered) available to the general public. The *next best* national commodity will be data; many practitioners refer to it as the *new oil*. Terms such as *data economy* will be increasingly common in coming years. Countries' economies will be partly measured and driven by the amounts of crude data that they own (the terms crude & raw are used to equate data with oil). Governmental data will be driving progress, friendships, and possibly conflicts and wars around the world in the near future. Governments that open their data to their citizens will benefit the most from that data, and will have the ability to being *data-empowered* governments. By sharing its data, government enables third-party academics, journalists, and many others to robustly adapt data in many new ways and find unexpected insights that are collectively beneficial to the country. However, to establish that, a number of issues need to be investigated, automated, and fixed, and that is exactly the goal of this book: to provide a breakdown, a step-by-step to taking the next leap at government, move into the age of data science and AI, and deliver the government of the future to the keen American public.

REFERENCES

A Report by IBM, 2007. The Hidden Costs of Data Migration – Strategies for Reducing Costs and Quickly Achieving Return on Investment. Published online.

Bertota, J., Gorham, U., Jaegera, S., Choiba, H., 2014. Big data, open government and e-government: issues, policies and recommendations. Information Polity 19, 5–16. http://dx.doi.org/10.3233/IP-140328. IOS Press.

City of Chicago Data Portal. Available from: http://data.cityofchicago.org.

IT Dashboard Federal Portal, 2016. Available from: https://itdashboard.gov/.

Joseph, R., Johnson, N., 2013. Leveraging big data: big data and transformational government. IT Professional 15 (6), 43–48.

National Institute of Standards and Technology, 2012. The Economic Impacts of Inadequate Infrastructure for Software Testing. US Department of Commerce.

San Francisco Governmental Data Website. Available from: http://data.sfgov.org.

United States Department of Agriculture's website. Available from: www.usda.gov.

Whitlock, C., Woodward, B., 2016. Pentagon Buries Evidence of $125 Billion in Bureaucratic Waste. A McKinsey and Company. Report published by the Washington Post.

Wikileaks Reports, 2017. Available from: https://wikileaks.org/ciav7p1/cms/index.html.

Yu, H.M.-T., 2012. Designing Software to Shape Open Government Policy, A Dissertation Presented to the Faculty of Princeton University in Candidacy for the Degree of Doctor of Philosophy. The Department of Computer Science, Princeton University.

Chapter 2

Disseminating Government Data Effectively in the Age of Open Data

Mirvat Sewadeh[1], **Jeffrey Sisson**[2]
[1]US Department of Agriculture, Washington, DC, United States; [2]US Census Bureau, Washington, DC, United States

Publicly available statistics from government agencies that are credible, relevant, accurate, and timely are essential for policy makers, individuals, households, businesses, academic institutions, and other organizations to make informed decisions. Even more, the effective operation of a democratic system of government depends on the unhindered flow of statistical information to its citizens.
Principles and Practices for a Federal Statistical Agency, Fifth Edition.

1. DATA DISSEMINATION IN THE FEDERAL GOVERNMENT: FROM COLONIAL AMERICA TO OPEN DATA

Dissemination of US official data, defined as "distribution or transmission of information or statistical facts by a public agency to the public" is an old practice that can be traced back to codification of laws in Colonial America. At that time, the rapid growth in the body of laws, and the limited availability of information about these laws, generated confusion and frustration among citizens and legislators and thus compelled lawmakers to start disseminating information about these laws (Tauberer, 2014).

But the dissemination of data—narrowly defined as statistical facts—by the US government may have well started with the release of the results of the first census, which took place in August 1790 (Colonial Times, 1790).

Since then, the collection and dissemination of data by the US government expanded to cover a large number of issues, ranging from economic activities and environmental conditions, to health and education. By the end of 2016, there were a total of 125 agencies whose work involves data collection or dissemination, of which 13 are principal statistical agencies that have statistical

Federal Data Science. http://dx.doi.org/10.1016/B978-0-12-812443-7.00002-8

activities as part of their core mission. The 13 federal statistical agencies are: the Bureau of Economic Analysis (Department of Commerce); the Bureau of Justice Statistics (Department of Justice); the Bureau of Labor Statistics (Department of Labor); the Bureau of Transportation Statistics (Department of Transportation); the Census Bureau (Department of Commerce); the Economic Research Service (Department of Agriculture); the Energy Information Administration (Department of Energy); the National Agricultural Statistics Service (Department of Agriculture); the National Center for Education Statistics (Department of Education); the National Center for Health Statistics (Department of Health and Human Services); the National Center for Science and Engineering Statistics (National Science Foundation); the Office of Research, Evaluation, and Statistics (Social Security Administration); the Statistics of Income Division (Department of the Treasury); the Microeconomic Surveys Unit, (Board of Directors of the Federal Reserve System); the Center for Behavioral Health Statistics and Quality, Substance Abuse and Mental Health Services Administration (Department of Health and Human Services); and the National Animal Health Monitoring System, Animal and Plant Health Inspection Service (Department of Agriculture).

The US Government views statistical data as a public good that provides critical information to public and private decision makers and the public in general. The collection and dissemination of statistical data by federal agencies are primarily governed by three statutes: The Paperwork Reduction Act, The Information or Data Quality Act (Data Quality Act, 2001), and the Confidential Information Protection and Statistical Efficiency Act (Information Protection and Statistical Efficiency Act, 2002).

In addition, several Statistical Policy Directives issued by the Office of Management and Budget (OMB) set minimum requirements for the principle federal statistical agencies and ensure the quality and coordination of federal official statistics. Of particular relevance to data dissemination is OMB's Statistical Policy Directive No. 4: Release and Dissemination of Statistical Products Produced by Federal Statistical Agencies; which was issued in March 2008 and sets principles, standards, and guidelines for the presentation and dissemination of statistical information. Among other things, the Directive requires agencies to ensure "equitable and timely access to data that are disseminated to the public" and to release documentation on statistical methodology and statistical frameworks that make it possible for users to verify data validity.

2. OPEN DATA POLICY: A NEW ERA IN DATA DISSEMINATION

The term "open data" refers to the idea that data can or should be made available for free use, reuse, and distribution. Advocates of "open data" emphasize the value that would be generated by making data available for sharing and redistribution to various users (Gurin, 2014; Open Data for Development Network, 2016).

With the US Federal Government going online in the late 1990s, there was a growing interest, both within the government and among the public, in making government-generated data publically available on the Internet in electronic formats. That interest culminated in the 2013 presidential executive order, which, along with a memorandum from the OMB, established the US Government's Open Data Policy. The Policy, which is part of a broad effort to promote the principles of Open Government, required agencies to ensure that new information collected or created supports "interoperability between information systems and dissemination of information to the public, without the need for costly retrofitting." To that end, agencies were required to use machine-readable and open formats for information as it is collected or created and use standards to promote data interoperability and openness.

The Open Data Policy presented a unique opportunity to modernize the collection and dissemination of data across the federal government. But it also placed federal statistical agencies in a demanding environment that is being reshaped by rapidly changing technology and evolving user expectations. More specifically, three broad trends have shaped the context of data dissemination:

Trend 1—Increased Demand for Statistical Information: The declining cost of sensors over the past decade, together with expanding storage capabilities, has allowed continuous gathering and storage of increasingly more granular data. At the same time, the shift of the World Wide Web to Web 3.0, which moved the emphasis of reasoning to data, has facilitated data reuse, and reduced the amount of human processing, allowing the release of large volumes of information and data that are currently inaccessible by making them machine processable (Bates, 2011). The increasing availability of data has fueled public interest in accessing, tracking, and analyzing data, generating demand for even more data. Whether it is policy makers using data to help formulate policies, scientists and academics interested in data to advance research, or businesses seeking to leverage information about consumers, users' demand for official data, and data in general, has grown exponentially. Access to data released by the US Census Bureau, through one of the agency's main dissemination systems, for example, has grown by more than 25% per year for the last 5 years. At the same time, the growing importance of data as a driver of private and public decision making has prompted users to look more critically at statistical information and increased their emphasis on accuracy, completeness, and consistency of data released by federal agencies.

Trend 2—Rising Importance of User-Focused Platforms: Technological advances have allowed data retrieval systems to become much more capable of identifying and selecting relevant material from large collections of data in response to users' queries. As a result, the dissemination of data has moved to user-focused and interactive platforms where users can easily find, download, manipulate, use, and reuse the data without restrictions. Interactive graphing and mapping tools have also become increasingly common on data dissemination platforms.

Trend 3—Evolving User Expectations: As a result of technological developments, consumers' expectations are also changing. New expectations require data providers to be ready to deliver and receive digital information anytime, anywhere, and on any device. Use of mobile devices to access data, for example, has risen significantly over the last 5 years, with some federal agencies reporting up to one-third of their users accessing data on mobile devices. And although user needs and preferences vary across user groups and individuals, users generally expect to easily find, download, manipulate, use, and reuse publicly available data without restrictions. Users also expect to be able to search and retrieve data by using plain language searches, rather than understand the structure of data sets or how data are classified. Recent use trends also show that data users are increasingly interested in interactive features, visualizations, and mapping of data, which allow the presentation of a large amount of information and statistical facts in easily accessible and user-friendly formats.

3. NEW ERA, NEW CHALLENGES

Responding to the growing demand for data amid a rapidly changing technological environment often requires federal agencies to establish new systems, reengineer or abandon existing ones, set up new processes, and develop modern platforms for delivering data (Refer to Box 2.1: Case study from Center for Disease Control).

Changes to data dissemination systems raise many challenges for federal agencies engaged in data dissemination activities:

Challenge 1—Data Silos: Chief among these challenges is the prevalence of "data system silos" across government agencies and within individual statistical agencies, which has led to the use of different approaches to the development of data products and dissemination systems. Some data silos date back to the times when official data were published only in print, and data integration was not possible for the end user. More recently, data system silos emerged as agencies, or individual programs within an agency, made "isolated" decisions on how to best create, package, and disseminate certain data. As a result, many federal agencies, or program areas within a single agency, developed unique data rules, data sets, processes, and procedures to serve specific needs with little, if any, coordination to maximize data congruency. In many cases, the fragmentation of data systems also led to the development of access tools that supported only a few, or even a single, data set.

The prevalence of data silos has many implications for data dissemination efforts across the federal government (Box 2.2). It often leads to inconsistent metadata (discussed in more detail later), variations in data definitions or aggregation levels across data sets, and differing processes and methods for expressing data quality. Some examples of these differences include:

- **Data definition differences** in what is included in the data collected for a given topic or inconsistency in how data are aggregated. An example is one agency includes only wages earned in the definition of income, whereas another agency includes capital gains, supplemental income from government programs, and other types of income in their definition.

Box 2.1 Health Data Interactive: A Data Tool With Decreasing Utility

Kate M. Brett[1]

The National Center for Health Statistics

The National Center for Health Statistics (NCHS) provides statistical information that is used to guide actions and policies to improve the health of the American people.

In the late 1990s, the NCHS began a data dissemination project using an off-the-shelf software designed to create customizable data tables using pretabulated data. At the time, the software offered new functions, including allowing users to rearrange rows and columns, nest additional characteristics within rows or columns, and hide or show items as desired. The success of the application among users prompted two other groups within the same NCHS division to launch two data dissemination applications using the same software; one focused on the health of older adults and the other targeted asthma data.

In 2011, and in an effort to consolidate the three dissemination tools and reduce the cost of maintaining separate applications, the NCHS launched the Health Data Interactive (HDI) tool. The HDI tool allowed users to find information for many of the most common health risks and outcome measures. Data in all tables were tabulated by data year, age, race/ethnicity, and sex. Where the data allowed, geography, urban/rural status, and income were also included as descriptive variables. Users could create graphs using the date and download the data to Excel or CSV files for further analysis and presentation. But the tool was difficult to learn and, because it rendered tables with potentially millions of data points, it required a lot of resources to maintain. At the same time, newer data dissemination systems provided much more user-friendly tools that are easier to maintain.

As a result, in 2016, the agency decided to discontinue the HDI tool and explore more advanced data dissemination applications. The decision to terminate the HDI tool was justified in that the tool's interface was not intuitive, the effort to learn how to use the tool did not translate into learning other NCHS tools, and agency leadership wanted to support emerging technology. Although it is easy to keep systems simply because they are not broken, the best systems should continually provide more utility over time, stay current in data policy requirements, understand their users' needs, and integrate into the agency data and information technology architecture rather than exist as stand-alone tools.

[1] Kate M. Brett is Research Scientist Officer at the National Center for Health Statistics/Centers for Disease Control and Prevention.

Box 2.2 US Census Bureau, Moving Beyond Data Silos

Jeffrey Sisson

The Census Bureau is one of the largest providers of demographic and economic census and survey data in the federal statistical system. The Bureau completes over 60 censuses or surveys as part of its portfolio or on behalf of other agencies. The existence of this number of surveys often leads to the emergence of data silos, with significant implications for processing and dissemination of data for the Census Bureau. The Census Bureau currently has over 50 different dissemination tools or methods for end users to get access to data, ranging from large, sophisticated tools such as American FactFinder (the Census Bureau's major dissemination system, https://factfinder.census.gov) and the Bureau's application programming interface (API), to simple tools built to disseminate one data set in spreadsheet formats.

As the Census Bureau began efforts to ease the complexity of getting data for end users, the existence of data silos complicated these efforts. The first American FactFinder was designed to bring together the largest and most significant and popular data sets released by the Bureau and provide a single access point to users. However, the differences in all of these data sets due to the variations in definitions of variables and inconsistency in metadata meant that once the various data sets are linked to the American FactFinder interface, users had to select a single data set to ease the navigation experience.

Current efforts by the Census Bureau continue to improve users' experience. The second version of American FactFinder gave users the ability to search by plain language topics or by geographies. It also offered different navigational paths to satisfy different user groups and their needs (discussed in more detail later). However, the data silos still existed internally and there was significant behind-the-scenes processing and mapping of data to "hide" those silos from end users. The Census Bureau has recently started efforts to break down these silos as much as possible in conjunction with the development of a new data platform. Although this is a promising effort, it will take many years to reach the point where data and systems can be truly integrated and the majority of the data silos are gone. The effort is also mainly focused on the integration of data that are being collected moving forward. It would require a significant amount of effort and resources to integrate historic data, which is unlikely to occur under current budget constraints.

- **Variation in definitions of geographical definitions**, which can be seen in the way data are tagged to a geographic classification (e.g., rural vs. urban), or in the way physical boundaries are defined. For instance, the US Department of Agriculture (USDA) definition of "rural" areas is based on programmatic needs such as the school lunch program, whereas that of the Census Bureau uses a geographical measure.
- **Data quality**: Agencies and programs within agencies often differ in both the measure of error they publish (margins of error, coefficients of variance, etc.) and the methodology they use to calculate the measures. These differences can be confusing to end users, especially those that are not as familiar with survey methodologies.

Challenge 2—Lack of Metadata Standards: Another challenge impacting data dissemination efforts in the federal government is lack of standards governing metadata (i.e., statistical and other information describing the data). Metadata serves many important functions, including helping users locate the data, defining the content of the data, and allowing software and machines to store, exchange, and process information. Therefore, it is critical for sharing, querying, and understanding statistical data, especially in the context of the semantic web. But for metadata to be effective, it has to be structured and consistent across its various sources in the federal government. The inconsistency of metadata hinders Interoperability and Communication Technology systems (Box 2.3).

Box 2.3 Using Metadata to Modernize Operations at the Bureau of Justice Statistics

Timothy Kearley[2]

The mission of the Bureau of Justice Statistics (BJS) is to inform criminal justice policy and practice by disseminating accurate, timely, and relevant information on crime and the administration of justice. BJS maintains over three dozen major statistical series designed to cover every stage of the American criminal and civil justice system. BJS disseminates the statistical information produced in the form of publications, data analysis tools, data tables, archived microdata, and related documentation.

BJS must improve public access to an increasing volume of data and a growing public demand for information in a variety of new formats. This must be done while adhering to strict data quality and security protocols that ensure that we are seen as a trusted, authoritative source of official information. As a result, BJS is gradually shifting away from the management of files and documents toward the development of centralized databases that store agency data, metadata, and knowledge at a granular level. One of the foundational components of this modernization effort has been to establish a consistent and comprehensive approach to managing our metadata.

Establishing a Metadata-Driven Architecture

The cross-cutting nature of metadata creates interdependencies across an organization. For example, the same metadata used to ensure coherence and consistency when information is exchanged between BJS and an external facility may also be used within metadata-driven dissemination tools to allow end users to reliably search, navigate, and interpret that information on our website. The same XML-based metadata used to describe a BJS questionnaire can be accessed to automatically produce online forms used to administer that questionnaire.

A few of the key challenges driving the transition to a metadata-driven architecture at BJS are presented in the following discussion along with related solutions that leverage metadata:

Challenge: Need to securely store and manage a large volume of data with variant attributes and security requirements.

Continued

[2] Timothy Kealey is Chief of Technology and Data Management at the Bureau of Justice Statistics.

Box 2.3 Using Metadata to Modernize Operations at the Bureau of Justice Statistics—cont'd

Solutions: Transition to a single database of record for all official statistics and maintain an accurate, connected metadata repository that includes information about data security requirements. BJS can then quickly respond to data calls, change numbers in one place and have those changes apply to many downstream access points, and apply new data security controls across all collections at once.

Challenge: Need to quickly produce and disseminate an array of products and data with limited staff.

Solutions: Establish tools and processes that use metadata to automate transformations and the production of reports, tables, and products. Publish products to the web on demand from the original underlying source data. Within online data access tools, use metadata about population sizes and aggregation levels to automatically suppress access when confidentiality protection thresholds are surpassed by user selections.

Challenge: Need to make it easier for internal analysts, external users, and machines to find, understand, and access information in multiple formats.

Solutions: Tag online content using a taxonomy drawn directly from the collection metadata to allow more accurate search and discovery for users and other systems. Use metadata and data virtualization tools to generate data cubes and longitudinal data sets that enhance data mining and visualization capabilities for analysts.

Many organizational changes are required to establish and sustain the use of a modernized, metadata-driven architecture. The work required cannot be taken on all at once and must be sequenced to account for project interdependencies. BJS is adopting a "service-oriented" approach that makes use of existing technologies and capabilities whenever possible as modernization efforts proceed.

Some of the key changes being undertaken include the following:

- Adopt established metadata standards and best practices
- Assign and define key roles (e.g., data stewards and curators)
- Establish standard naming conventions and business rules
- Develop and maintain a reference metadata bank at the agency level that defines organizational units, products, variables, concepts, classifications, and controlled vocabularies
- Implement new technologies for data and metadata management
- Establish methods and systems for assigning unique identifiers
- Establish a collaborative space for sharing and accessing all this information
- Set new policies for metadata management and data processing
- Train staff to gradually build up in-house capability and capacity
- Automate data processing and publishing processes

By establishing a metadata-driven architecture and tools, automation will be used to increase efficiency. Access to information for both our internal and external data users will be enhanced by ensuring that data are easier to find and can be presented in whatever manner is most useful to a broad audience (Fig. 2.1).

Box 2.3 Using Metadata to Modernize Operations at the Bureau of Justice Statistics—cont'd

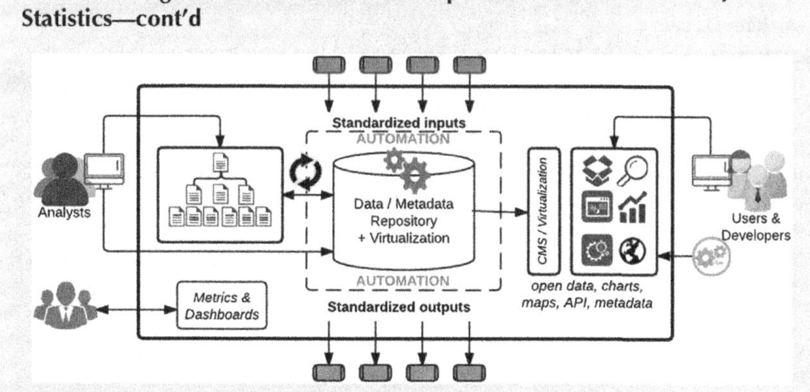

FIGURE 2.1 Metadata-driven enterprise architecture. *From Bureau of Justice Statistics (BJS).*

Challenge 3—Understanding User Needs: As discussed earlier, expectations of data users are constantly evolving with changes in technology. Users expect to get data and information tailored to their needs and to be able to access that information seamlessly on any device at any time. Federal agencies providing statistical and other data have to overcome the challenges of both understanding customer needs and responding to the ongoing changes in these needs and expectations. For many agencies, this requires a shift in "culture" to meet users' needs with relation to the type of data release and dissemination methods. For many agencies this shift could mean implementing new procedures and processes to capture feedback from users and incorporate this input into decision on data product changes.

Challenge 4—Communicating Uncertainty. Data errors can occur during the collection or processing of data, arise from errors in sampling or coverage, or arise from nonresponse, among other sources. In addition, there is an inherent uncertainty that occurs solely as a result of using a sample from a population, instead of conducting a census (complete enumeration) of the population in question. US federal agencies often report official statistics as point estimates, without providing measures of error. As a result, users of the data may incorrectly view these statistics as error-free or may incorrectly underestimate error magnitudes (Manski, 2015). Communicating uncertainty about official data is particularly important given the impact that data released by the government can have on markets, policy making, and the public in general. Yet different users vary in their understanding of measures of uncertainty and, therefore, it can be difficult for federal agencies to effectively educate end users on how to interpret or appropriately use the data. Novice users tend to be less informed about uncertainty in data and are more likely to ignore any information provided about the measures of error around an estimate. However, more advanced users relying on data to make critical decisions need to understand and make informed decisions on the use of the data and any associated measures of uncertainty (Refer to Box 2.4: Bureau of Labor Statistics responding to user needs and communicating uncertainty in data).

Box 2.4 Communicating Uncertainty in Official Statistics

Michael D. Levi[3]

The Bureau of Labor Statistics (BLS) of the US Department of Labor is the principal federal agency responsible for measuring labor market activity, working conditions, and price changes in the US economy. Maintaining the credibility of BLS' published data is essential to the Bureau's mission and critical to data users. However, survey results come with some uncertainty and data users need to understand both the strengths and limitations of the statistical information they use. Although BLS has long published standard errors for many of our surveys, this information was somewhat obscure, often hidden in technical documentation, dense tables, or available only upon request.

In recent years, in an effort enhance transparency and contribute to the statistical literacy of data users, BLS launched an initiative to better explain and display measures of survey reliability. One outcome of this initiative was a series of posts in the BLS "Commissioner's Corner" blog. The blog uses two BLS examples to explain, in nontechnical terms, what is meant by sampling error and how to understand confidence intervals.

Another outcome of this initiative was visual displays of survey estimates and their accompanying confidence intervals. The visualization effort began with a session for the BLS Data Users Advisory Committee, which represents a diverse group of users and provides advice to BLS on matters related to the analysis, dissemination, and use of the Bureau's data products.

The Current Employment Statistics (CES) program presented several possible charts employing a variety of visual techniques to convey the 90% confidence interval surrounding estimates of nonfarm payroll data, one of the most prominent data sets produced by the BLS, and asked for feedback. The committee selected an approach and recommended that the program reach out to the public for additional comments. The CES program refined the preferred approach into two charts and conducted an online survey that included an explanation of the data, clarifying text about each chart, and eight questions about the charts. The survey, which targeted existing CES data users and had a goal of 600 responses, received almost three times as many responses.

The majority of online respondents, 82%, found the charts useful in identifying statistically significant changes more quickly and getting a better sense of how the variability of estimates differs among industry sectors. Responses from users that self-identified as basic, intermediate, and super users were not substantially different and were overwhelmingly in favor of publishing the chart with the data.

Based on the results of the survey, the CES further refined the visualization and, in December, 2016, BLS began publishing an interactive chart showing 1-month, 3-month, 6-month, and 12-month changes in employment by industry with confidence intervals as a standard accompaniment to its monthly Employment Situation news release (Fig. 2.2).

---

[3] Michael D. Levi is Associate Commissioner for Publications and Special Studies at the Bureau of Labor Statistics.

Box 2.4 Communicating Uncertainty in Official Statistics—cont'd

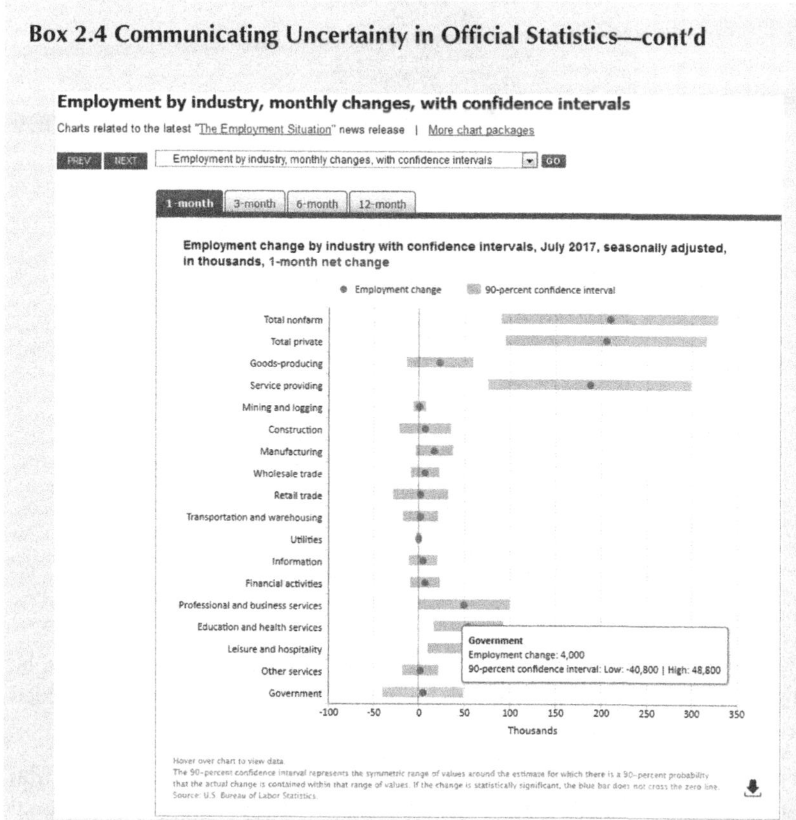

FIGURE 2.2 Bureau of Labor Statistics visualization example. *https://www.bls.gov/charts/employment-situation/otm-employment-change-by-industry-confidence-intervals.htm.*

4. TOWARD A MODERN AND EFFECTIVE DATA DISSEMINATION STRATEGY

An effective data dissemination strategy should be designed to provide users with the data they need in the right format and in a timely fashion, while ensuring accuracy and integrity of the data. Put differently, the focus of a successful data dissemination paradigm should be the users' needs, wants, and constraints with relation to the content of the data, the format in which data are released, timeliness of delivery, and the technology used to deliver the data (Fig. 2.3).

In developing a user-focused approach to data dissemination, it is important to recognize that user needs and expectations are not only constantly changing in response to technological advancements, but also vary significantly across user groups. To respond to various user groups, federal agencies engaged in data dissemination can use different approaches to communicate data to reach different users (Speyer et al., 2014).

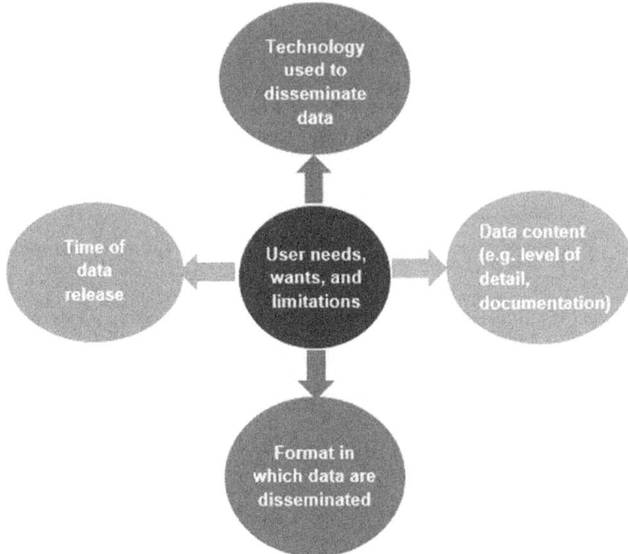

FIGURE 2.3 A user-focused data dissemination strategy.

To help understand and address the needs of various data users, it would be useful to divide users into three major groups based on their needs and intended uses of the data:

- **General public**: The largest number of data users are made up of users who are not subject matter experts (e.g., students working on term paper or casual users interested in obtaining some data for one-time use). Members of this group tend to be interested in single point estimates or mega trends over time. They are generally less experienced data users and struggle with complex dissemination. This group needs simplified paths with plain language guidance to get them to a limited set of data for their topic of interest. Punchy data and visualizations showing key trends are also very effective in communicating data to this group. A good example is a recent effort by the Economic Research Service (ERS) to modernize the delivery of some of its data. In March 2015, the ERS began releasing forecasts on US agricultural trade, the fruit and vegetables markets in a series of interactive charts and maps. The agency used a new visualization software to make technical information and data—traditionally available in long reports released in PDF format—available in a more modern, user-friendly, and easily accessible format.
- **Media and policy makers**: The second largest group of users comprises members of the media and policy makers in general. This group has more specialized knowledge of the topic in question and tends to be interested in statistics that have relevance to specific policy issue or "story." For this group, data visualizations and briefs describing key findings can be very effective. A

good example of data dissemination approaches targeted at policy makers and media is the Charts of Notes series, which is released daily by the USDA's ERS and provides charts drawn from the agency's reports and data products containing notable facts and figures about the US agricultural sector. Media and policy makers are more adept at navigating complex search and download tools, but still prefer simpler methods of getting to the larger quantity of data they need. The media, especially, needs simplified methods as they are often working on stories under tight deadlines and do not have time to search and analyze data. Media and policy makers are also very interested in map-based data that can be easily incorporated into stories or briefs.

- **Data analysts and academics**: This is perhaps the smallest group of users in terms of numbers, but the one that requires the most complete data sets and has the most need for information about data collection and processing methods. They tend to use query tools and APIs and require documentation regarding methodology and other technical information. To meet the needs of this group, many federal agencies have developed one or more APIs to deliver data in machine readable format to data analysts, academics, and others who use the data in analytical models or scientific studies.

To meet the ever-changing needs of data users and satisfy user groups with different needs and levels of understanding, agencies will need to develop dissemination platforms that are agile and can adapt quickly. Technical advances can help drive this process, but are hindered by the existence of data silos mentioned earlier. As agencies begin to break down the data silos, they can create integrated and standardized data and metadata repositories that will be the basis for data access moving forward. These integrated repositories will allow agencies to create "lightweight" data access tools that can be customized to user groups, access methods (mobile vs. desktop) or data displays (tabular, map-based, visualizations, etc.).

A common data and metadata repository would also allow agencies to publish data once and create tools that access the single instance of data and metadata. The paradigm of "publish once, use many times" is critical in moving agencies toward an agile, user-centric data dissemination approach and ensuring data quality. Having a single instance allows agencies to ensure that data changes and corrections are handled efficiently and not be concerned with updating multiple instances of the data. It would also allow the areas creating the data for publication to have a single delivery point using a standard format. These factors are all critical in enabling federal agencies to develop a more modern dissemination platform under the current budget and resource constraints.

The breakdown of these data and metadata silos needs not only to begin within each agency, but also to spread and be coordinated between agencies. The federal government has recognized this need through publishing guidance on open data and interoperability between agencies, but significantly more work needs to be done. The demand for integrated data will continue to grow as the

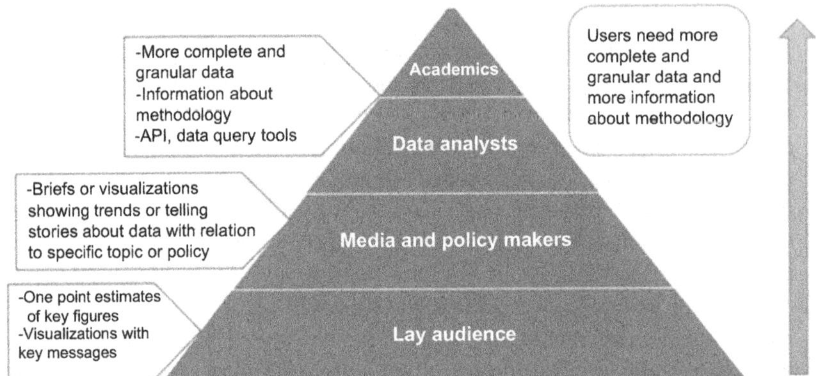

FIGURE 2.4 Data user segments and needs.

push to make data-driven decisions increase and federal agencies, especially the principle statistical agencies, are in a unique position to provide the foundation for such integration. The creation of an agile dissemination platform will give agencies the ability to provide end-user-focused data access, but it does not ensure that this will happen (Fig. 2.4).

At the same time, to better understand users' needs and wants, agencies engaged in data dissemination should seek continuous feedback from users. One way to obtain such feedback is by adopting a Customer Experience Management (CEM) system to ensure that customer interactions are captured and assessed systematically and that information obtained is used to further refine data products and dissemination platforms. Agencies should use web analytics systemically to collect, measure, and analyze the usage of data. Many federal agencies also use online surveys to gather information from users about their online experience.

Agencies also need to continue to find creative ways to engage with their data users earlier in the development of new products and systems. Gathering user feedback early in the development process will allow government agencies to test ideas without spending much time and resources developing a tool that will not meet customer needs. Some agencies have created groups of users to provide early feedback on ideas for new data products or early testing of new tools. Such groups allow an agency to get early feedback on innovative ideas without investing significant resources in the development of a new product or system that may not meet user needs.

Finally, ensuring data accuracy is perhaps the single most important component of any data collection and dissemination program. No matter how easily users can access data or how modern the mode of the dissemination platform is, if users do not have confidence in the integrity and accuracy of the data being released, they will not use it. As mentioned previously, data errors can occur at different stages during the collection, analysis, or processing of data. Ensuring data quality requires developing a system for monitoring various sources of errors and implementing strategies for reducing errors during the estimation, measurement, or processing of data. Clearly, many of these strategies are implemented long before the data are disseminated to the public. But any dissemination system should include three quality assurance components: a process or system for data validation, guidelines on communicating uncertainty about data to users, and a process for informing users about errors identified in data after they were released.

The objective of data validation during the dissemination stage is simply to profile the data for missing data points, inconsistencies, or outliers. The verification process could be carried out by an individual who is familiar with the data. Alternatively, agencies can use a simple automated process that identifies errors. The ERS, for example, implements a tool that identifies and flags data outliers where data points that are one or two standard deviations from the mean are flagged automatically.

5. CONCLUSION

Demand for and use of US Government data will continue to grow in the foreseen future, driven by technological change and rising interest in more and increasingly granular data. This places increasing pressure on government agencies to expand access to official data, while ensuring its timeliness and accuracy.

For federal agencies involved in disseminating official data, the Open Data policy has served as a catalyst for modernizing their data collection systems and dissemination platforms to meet users' demands and expectations. As shown in the examples drawn from some federal statistical agencies and discussed in the chapter, the scale and scope of these modernization efforts varied depending on agency priorities and objectives.

However, more work remains. As federal agencies continue their modernization efforts or embark on new ones, it is important to improve coordination among government agencies to break down data silos and minimize the fragmentation of the federal data system, while also integrating new technology to ease the transition. New search and navigation techniques can mitigate existing data silos and get end users to desired results more quickly. At the same time, new mapping and visualization tools can facilitate the integration and interpretation of disparate data for end users.

And to meet the diverse and evolving needs of data users, agencies should aim to develop agile data dissemination platforms that can adapt more quickly than they have historically. Federal agencies would also benefit from gathering continuous feedback from users and disseminating data in different formats.

DISCLAIMER

The views expressed are those of the authors and should not be attributed to the ERS, USDA, or US Census Bureau.

REFERENCES

Bates, M., December 2011. Understanding Information Retrieval Systems: Management, Types, and Standards. CRC Press, Boca Raton, Florida.

Gurin, J., 2014. Open Data Now, the Secret to Hot Startups, Smart Investing, Savvy Marketing, and Fast Innovation. McGraw-Hill.

Manski, C.F., 2015. Communicating uncertainty in official economic statistics: an appraisal fifty years after Morgenstern. Journal of Economic Literature 53 (3), 1–23.

Open Data for Development Network, 2016. Open Data for Development; Building an Inclusive Data Revolution Annual Report Available from: http://od4d.com/wp-content/uploads/2016/06/OD4D_annual_report_2015.pdf.

Paperwork Reduction Act of 1995, 1995. Pub. L. No. 96-511, 94 Stat. 2812, codified at 44 U.S.C. §§ 3501–3521.

Speyer, P., Pagels, B., Ho-Park, N., 2014. Communicating Data for Impact. Forum One.

Tauberer, J., 2014. Open Government Data, second ed. Available from: https://opengovdata.io/2014/ancient-origins-open-access-to-law.

The Confidential Information Protection and Statistical Efficiency Act, 2002. Pub. L. 107-347, 116 Stat. 2899, 44 U.S.C. § 101.

The Federal Data Quality Act, 2001. Pub. L. No. 106-554—C 114 STAT. 2763A–153.

U.S. Department of Commerce, Bureau of the Census, 1975. Historical Statistics of the United States, 1790, Colonial Times to 1790 Part 1. Government Printing Office.

Chapter 3

Machine Learning for the Government: Challenges and Statistical Difficulties

Samuel Eisenberg
Jersey City, NJ, United States

The difficulty lies, not in the new ideas, but in escaping from the old ones.
John Keynes (1939)

1. INTRODUCTION

Humanity, civilizations, and organizations are generating and storing ubiquitous amounts of data, and they are growing at a tremendous rate (NSA Utah data center, 2017). The hardware and toolsets required to handle this size of data have become capable and readily available, and cheap. Data analytics, data mining, automated decisions, and machine learning are not new tools to governments and businesses (Samuel, 1959). These tools have been leveraged for years on specific problems, such as cost forecasting (GAO Cost Estimating, 2016). Before the information age of computers, there have been teams of mathematicians, statisticians, and engineers who optimized processes and added levels of mathematical rigor. What is happening now is different. It is at the point where the automation of analytical capabilities brings any single analyst with tooling to levels outperforming dozens to hundreds of statisticians and accountants. This move from human to automated labor is identical to how it used to require many people to handle the routing of phone calls or the transcoding of telegrams, also paralleling the changes to the manufacturing industry as robotics was introduced. The manual analysis efforts are left further and further behind as the volume of data increases. Although widespread prolific networks and powerful computers have unlocked potential, there are still difficulties in adapting and revitalizing processes to use business intelligence and machine learning.

The government faces many challenges as automated analytics are introduced and ingrained into decision making and process. A strikingly significant amount of

Federal Data Science. http://dx.doi.org/10.1016/B978-0-12-812443-7.00003-X

data entry and information sharing still relies and operates via spreadsheets (USA Jobs, 2017). With new techniques, incorrect solutions will often yield poorer results than the existing processes in place. Mistakes are not as apparent as the process is hands off, and automation can have the wrong solution extensively deployed while manual efforts are limited in scope. Imagine one bad switch board operator mangling a few phone calls versus an algorithm, which if incorrectly deployed could drop all phone calls. Change can be costly and risky. Despite that, projects are easier to implement and solutions are increasing in sophistication. The challenges and solutions to correctly utilize data analytics are becoming well known and the tools are now abundant. There is continuing development at the forefront of automated intelligence and sophisticated techniques but the biggest impact is coming into fruition with mass deployment and utilization. Change control and education separate the leading edge from the practitioners, but the gap is shrinking.

2. AN INTRODUCTION TO DATA MINING

One of the most important steps in the data mining process is the stage where the analytical question or decision is composed of components of available data, known algorithms, and assumptions. Advanced techniques exist to catalog the available data; but, more often than not, manual data dictionaries suffice with succinct data warehouses. Data collection and cleaning can sometimes be the longest and costliest step of an analytical exercise (USA Census, 2010). It requires constant work and processes to maintain and should not be thought of as a one-time exercise (Data Governance, 2017). The reader will assume for the remainder of the chapter that the available data are at least labeled and known well, as to not get bogged down with data collecting and cleansing. Most data mining algorithms can be grouped into classes such as regression, classification, similarity, clustering, market basket, profiling, data reduction, link and node prediction, and causal modeling, based upon which common data solution is being employed. Most of these solutions all share basic statistical underpinnings, so instead of jumping into a rigorous definition of one or fully surveying the breath of currently solvable problem types, an exploration of a few basic concepts at a high level is provided to build a sense of familiarity.

Regression is one of the most well-known and widespread techniques, intuitive and mostly well behaved. It is robust against the common issues outlined later in this section and has very natural tangible auditing techniques such as graphical visualizations. It could be considered as the class of algorithms where an estimated value is created, most often for the use of prediction and forecasting. Linear regression is one of the simplest effective regression modeling tools (and Microsoft Excel certainly has seemed to push its utilization) (Liebowitz and Margolis, 1999). Other forms include random forests and support vector machines (Encyclopedia of Mathematics, 2017) and are becoming more available and easier to use breaking the linear regressions dominance in common analytics.

2.1 Learning With Orange

For the remainder of this chapter, a general public licensed machine learning educational application named Orange3 and its included example data sets (Orange, 2017) will be shown. This tool was not chosen for its sophistication or its power; it would not be appropriate for the vast amount of government data problems. Orange is one of the simplest tools, visually programmed, and comes equipped with classic training data sets that are seen in multiple courses (Fig. 3.1). It has a very low learning curve and even comes with built in tutorials. This lets the reader get his/her hands involved with the process before continued reading and education that would offer a more robust introduction and experience. Just as children learn on tricycles and bicycles with training wheels, after a basic comfort or intuition is built, more powerful tools, such as R, Tensor Flow, Python, to name a few, are easier to pick up and adopt for real-world problem solving. For brevity, only the final setup along with a few images investigating the data and models will be given during the walk-through to allow the reader to follow the tool's included tutorial videos and then reconstruct the chapter examples and experience using the tool. The following is the complete yet simple setup used for demonstrating basic concepts.

The first step is connecting the data file to the data table so that one can view the full contents in a direct raw format. For this exercise, the included housing price tabular data file is used. This is done so that one can understand what type of data exists for our examples. This is often skipped within production analytics systems as the data gathering, cleaning, and labeling have already been performed by subject matter experts within the organization. After looking through

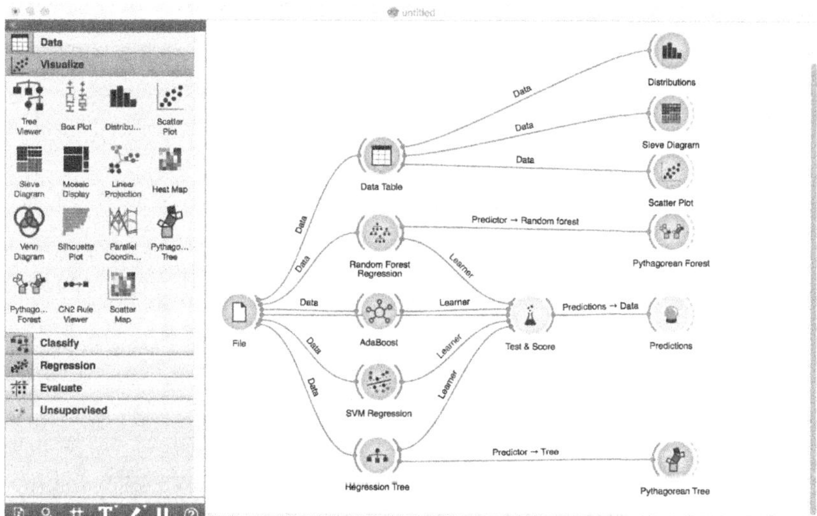

FIGURE 3.1 End state of the Orange housing.

the table to get an idea of what the data are, the reader should connect a few of the visualization tools such as scatter plots, distributions, or sieve diagrams to get a sense of the data distribution and correlations. This is performed as many algorithms have strengths and weaknesses around the types and distributions of the underlying data sets. This chapter makes no attempt at providing such an intuition; instead the ease of the tool is used with a brute force approach in that multiple methodologies are used on the training data.

The distribution of median home value verses housing density is shown. A quick visual inspection can show that a predominate Gaussian curve is seen with a potential smaller secondary one as an additive. Very early into the data investigation process, it is apparent that housing density over 0.01 can be used to accurately estimate the median housing value (Fig. 3.2).

The second visualization is more complex in that there are three variables in play. Distance and lower income ratios are compared with the median housing value. Intuitively, lower income correlated with lower home values and distance inversely relates to home value (Fig. 3.3).

In a similar visualization room count is switched with distance. The intuitive result of more rooms relating to a higher home value is seen (Fig. 3.4).

A sieve, or parquet, diagram shows the relationship between medium home value and the housing tax. The diagonal line with more example data points shows a tight correlation between housing value and housing tax. There are not many examples of low value with high tax or high value with low tax. This again is not very insightful, it is expected as the home value is often a large contributor toward the tax

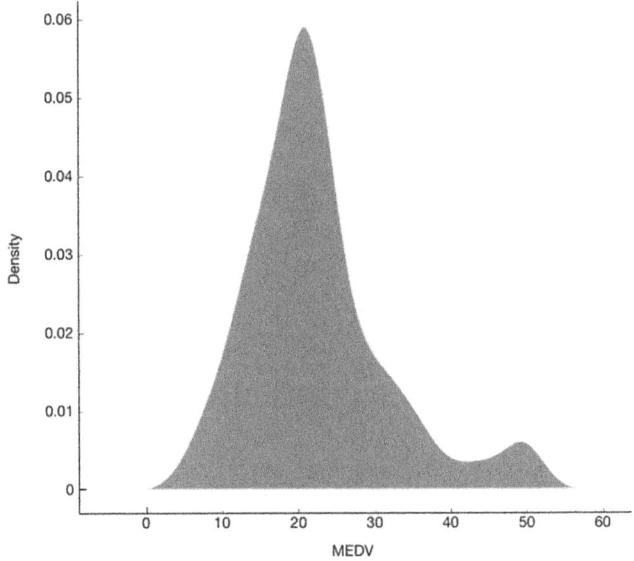

FIGURE 3.2 Median home value distribution.

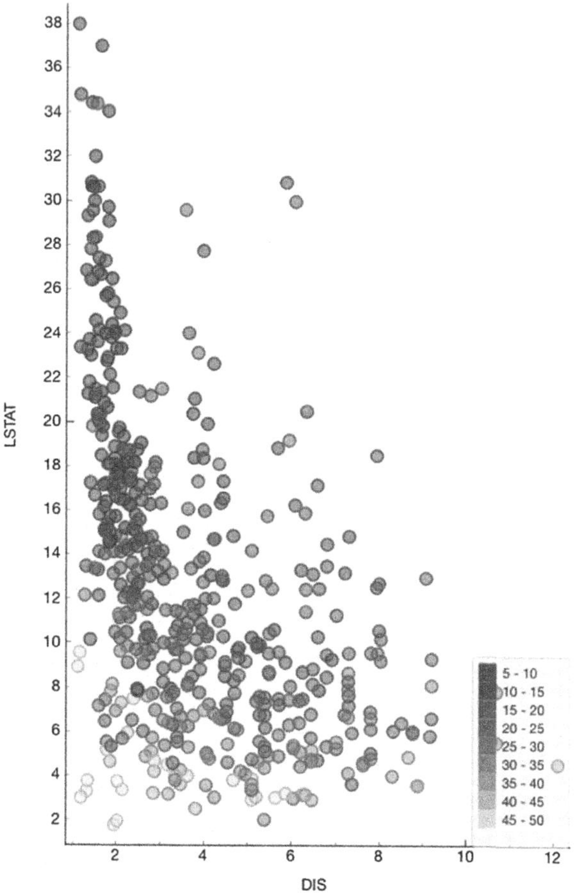

FIGURE 3.3 Distance and lower income to median home value.

due calculation. Finding a similar or reversed pattern with unexpected indicators or with one that is known before another could prove highly useful (Fig. 3.5).

One straightforward and relevant use of this demonstration file is to estimate the medium value of an unknown house. While the tax offices in a local community should know each house, a few civil engineers might need to estimate the values and start using them with assumptions and models. Connect the random forest regression, regression tree, support vector machine, and adaptive boosting (ensemble) widget to the data and test and score widget. These models all attempt to utilize features of the data to be able to make predictions of other data points. The full details of each model are outside the scope of this chapter, and each model should be thoroughly studied for a true understanding, but even without an understanding testing methods can let us have some trust in the results despite it being a black box. Connect the data to the test and score widget and feed the predictions to the predictions widget.

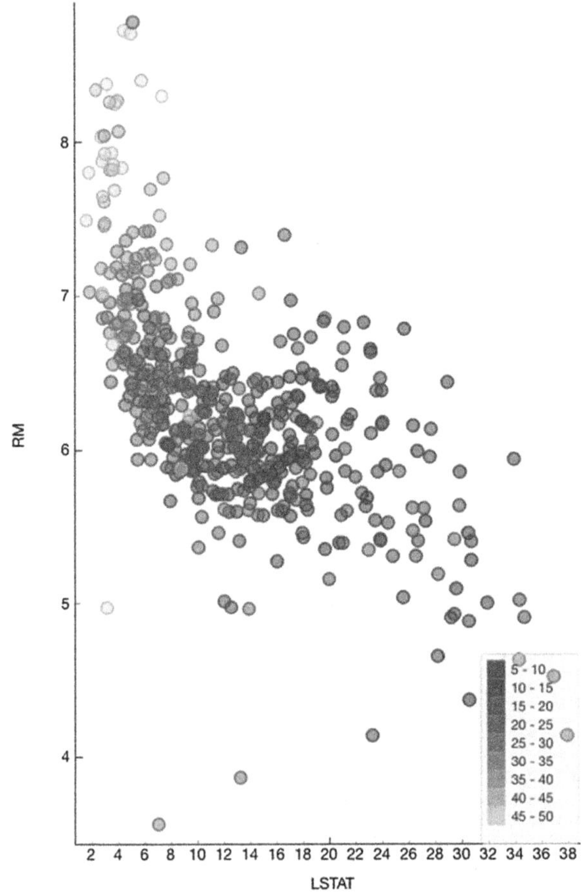

FIGURE 3.4 Room count and lower income to median home value.

The picture (Fig. 3.6) of the model predictions verses the actual values shows data point by data point how the models perform.

The graphical representation of the trees (Fig. 3.7) shows the depth of branching in the model. It intuitively shows how much each individual tree can vary but is not directly useful for trying to pull out generative insights where rules can be described as one would with some of the basic distribution visualizations.

3. RESULT VALIDATION, TRUST BUT VERIFY

At this point it is almost negligent to lack any form of validation with a model. A master who knows the distributions and models could potentially pull it off without hitting issues, but many individuals applying the techniques will not have that level of intimacy or experience. Despite that, the ability to test the

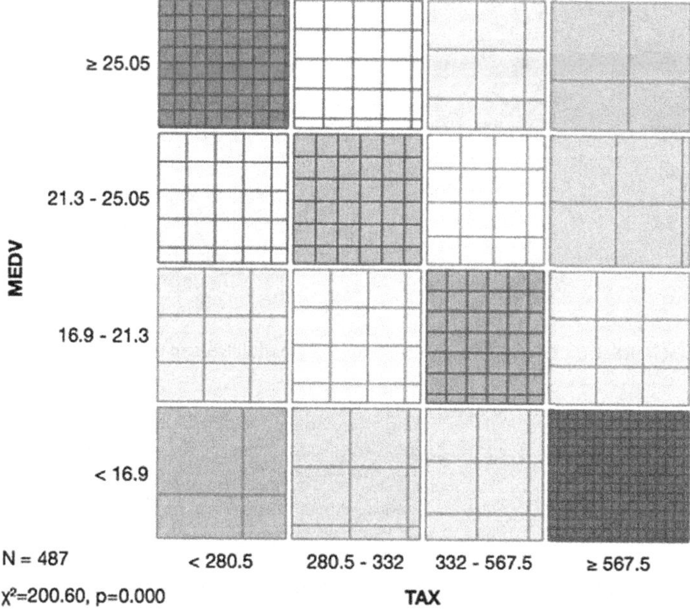

FIGURE 3.5 Tax to home value sieve.

Predictions

Info
Data: 506 instances.
Predictors: N/A
Task: N/A
Restore Original Order
Data View
Show full data set
Output
Original data
Predictions
Probabilities
Report

MEDV	dom Forest Regres	Regression Tree	SVM Regression	AdaBoost	Fold	CRIM	ZN	INDUS
24.000	30.288	32.300	29.429	27.100	1	0.006	18.000	2.310
21.600	21.916	22.300	23.152	22.900	2	0.027	0.000	7.070
34.700	34.562	33.450	31.112	33.400	3	0.027	0.000	7.070
33.400	35.429	35.150	28.872	34.900	4	0.032	0.000	2.180
36.200	34.326	33.125	29.223	35.100	5	0.069	0.000	2.180
28.700	24.156	24.550	23.833	24.000	6	0.030	0.000	2.180
22.900	21.727	17.300	21.446	19.800	7	0.088	12.500	7.870
27.100	16.424	16.900	19.252	17.800	8	0.145	12.500	7.870
16.500	17.173	16.950	16.865	15.000	9	0.211	12.500	7.870
18.900	21.210	16.300	19.014	18.900	10	0.170	12.500	7.870
15.000	23.745	17.700	20.002	22.200	11	0.225	12.500	7.870
18.900	20.095	19.625	20.314	19.800	12	0.117	12.500	7.870
21.700	21.241	22.100	20.944	19.500	13	0.094	12.500	7.870
20.400	20.101	19.975	19.393	19.900	14	0.630	0.000	8.140
18.200	20.899	20.675	18.598	20.300	15	0.638	0.000	8.140
19.900	20.804	20.275	19.264	20.400	16	0.627	0.000	8.140
23.100	21.816	22.450	21.311	20.600	17	1.054	0.000	8.140
17.500	18.008	18.050	17.094	18.400	18	0.784	0.000	8.140
20.200	19.455	19.150	18.528	18.200	19	0.803	0.000	8.140
18.200	19.763	19.767	17.965	19.700	20	0.726	0.000	8.140
13.600	15.469	15.050	14.798	14.500	21	1.252	0.000	8.140
19.600	19.535	18.475	17.026	19.600	22	0.852	0.000	8.140
15.200	14.986	17.250	16.666	17.800	23	1.232	0.000	8.140
14.500	15.079	13.700	15.288	15.000	24	0.988	0.000	8.140
15.600	17.728	17.833	16.221	17.400	25	0.750	0.000	8.140
13.900	15.405	16.200	15.333	16.600	26	0.841	0.000	8.140
16.600	17.374	16.650	15.857	17.500	27	0.672	0.000	8.140
14.800	14.396	15.525	16.071	15.600	28	0.956	0.000	8.140
18.400	20.839	18.900	19.298	19.600	29	0.773	0.000	8.140
21.000	22.202	23.750	20.704	20.300	30	1.002	0.000	8.140
12.700	14.296	15.725	14.670	13.900	31	1.131	0.000	8.140
14.500	20.324	19.500	17.256	19.800	32	1.355	0.000	8.140
13.200	14.906	18.150	15.348	14.000	33	1.388	0.000	8.140
13.100	14.427	15.900	15.288	13.900	34	1.152	0.000	8.140
13.500	13.561	13.250	[illegible]	19.600	35	1.613	0.000	8.140
18.900	21.237	24.500	20.685	21.000	36	0.064	0.000	5.960

FIGURE 3.6 Medium home value and predictions.

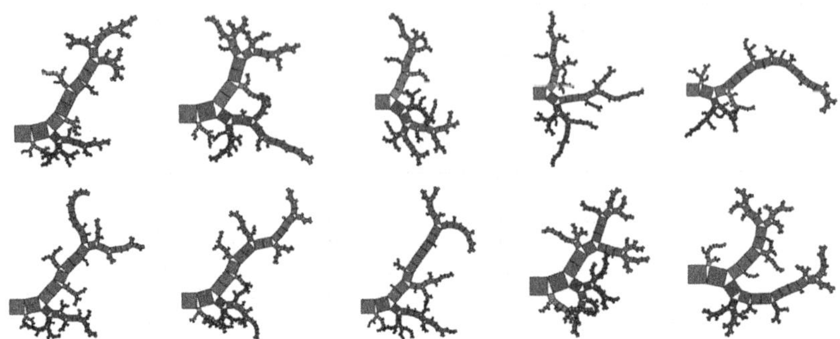

FIGURE 3.7 Graphical representation of regression trees in the random forest.

Test & Score

Sampling

Cross validation

Number of folds: 10

Stratified

Random sampling

Repeat train/test: 10

Training set size: 66 %

Stratified

Leave one out

Test on train data

Test on test data

Report

Evaluation Results

Method	MSE	RMSE	MAE	R2
Random Forest Regression	12.862	3.586	2.378	0.848
Regression Tree	19.472	4.413	3.049	0.769
SVM Regression	28.316	5.321	3.097	0.665
AdaBoost	10.658	3.265	2.119	0.874

FIGURE 3.8 Medium home value cross-validation.

models on data that were withheld from training can provide a certain level of trust in the results. This becomes more important as the techniques become more convoluted and harder to grasp as well as for adding protection for future scenarios were models are no longer functioning properly. Most libraries and tools contain easy-to-use cross-validations, confusion matrixes, random sampling, etc. to evaluate how well the prediction performs against data withheld from modeling. This is such a crucial step that the Orange tool set the testing and scoring widget to be responsible for the conversion of predictors into predicted values, so that you cannot use the model without having the testing built in. As a challenge, attempt creating a nearest neighbor classification on the Iris data set and evaluate it with the confusion matrix attached to its test and score widget to understand its performance and get an idea of how often one can anticipate it misclassifying a flower (Fig. 3.8).

3.1 Iris Aside

The Iris data set is one of the most widely used because of its simplicity, intuitiveness, and historic nature. It was introduced by Ronald Fisher through a research paper but was actually collected by Edgar Anderson (Fisher, 1936). It consists of 150 characteristic measurements from three classes of Iris flowers with labels. It can be used for both supervised and unsupervised learning techniques in which case the labels are considered or omitted, respectively.

4. MODEL OVERFITTING, TOO GOOD TO BE TRUE

It is trivial to create a model that perfectly categorizes the training set of data; it would just consist of the full training data so that it could look up the answer. This contrived model would have no actual value or predictive power. The real risk is not this absurd or easily recognized; models are slowly overfitted through iterations and slip past detection. It can naturally pose a risk to data sets and certain algorithms can have issues with it when outliers are included. Most of the tooling has built in protection from overfitting at the time of training by withholding information. Occasional validations, with alternative data sets from the same population throughout the lifespan of the solution act to ensure that models are not overly trained. Overly trained models can appear to be working well with the standard range of input; they are dangerous around the infrequently seen events. Depending upon the data task at hand dictates how cautions and vigilant one must be, and after more exposure one can develop a good intuition on when it poses a risk to the model.

5. STATISTICAL BIAS, IMPACTING RESULTS BEFORE ANALYSIS BEGINS

Statistical bias can be introduced into a model from a variety of sources. Often it can be accidentally introduced by the unwary selection impact an action has. It can also be frequently found where assumptions have become incorrect and not revalidated. Selection bias, survival bias, reporting bias, attrition bias, exclusion bias, etc., are a prevalent issue from multiple forms that can even break fundamental assumptions such as the central limit theorem. It can be a challenge to pull apart bias from skew, even when looking at individual distributions in the data. This chapter will not show a visual example, because bias can be introduced quite early on, and even validation techniques and holdout groups give it enough room to hide in. It is recommended to review the data collection process and assumptions to identify the actual cases of bias being introduced into results. If warranted, a side effort with different methodology can be conducted to ensure that data patterns seem to match, leaving bias much less likely to be involved.

6. SEGMENTATION AND SIMPSON'S PARADOX

Hidden segmentation can cause incorrect predictions when applied to whole populations; being unaware of, or incorrect on, a subtle assumption can completely reduce a solution to a useless or harmful one. But one cannot avoid segmenting; it is required and needed in analysis. In fact, one most often sees good analysis performed with highly aggregated data, the details can cause one to "not see the forest through the trees" (as a metaphorical expression, nothing to do with regressions that share those names). Simpson's paradox is a counterintuitive fact of summarized data where the trend is diminished or even reversed when groups are combined. This can be quiet startling when one is first introduced to this thought, but it quickly makes sense and can easily be demonstrated. This is a common result of a causal relationship or correlation in the data, often caused by a confounding variable naturally occurring from processes or distributions in place. It is best not to slice and dice everything to the point of exhaustion for checking, but it is something to be leery of and double check against before finalizing decisions.

There are multiple cases where Simpson's paradox has been found. An old but well-known example is the passage of the Civil Rights Act of 1964. It is an illustrative example of where the wrong understanding would be found by those unaware of the lurking geography component. With the voting detailed later, one sees that at the time Democrats voted with higher percentages for the act than the Republican counterparts within each region; however, the total Republican votes tallied shows a higher percentage for the act than the totaled Democrats tallied votes. Identified cases of Simpson's paradox often lead direct insights to a useful trend in the data; in this example, the issue is found to be much more prevalent along geographical locations instead of pure political ideologies (Fig. 3.9).

	Democrats	Republicans
Total Voted	152/248 (61%)	138/172 (80%)
Northern Voted	145/154 (94%)	138/162 (85%)
Southern Voted	7/94 (7%)	0/10 (0%)

FIGURE 3.9 A case of Simpson's paradox.

7. OUTLIERS AND BAD DATA

Outliers most often cause issue by reducing the effectiveness of certain algorithms, but they can also be a telltale sign that there is skewed data, multiple underlying distributions, or infrequent tail events. These can be quiet troubling because the only solutions are data completion, record removal, segmentation, and outlier robust statistics. There are algorithmic approaches to detecting or removing outliers, but they should never be applied blindly without a judgment call from investigating their presence. Incorrect removal of outliers can do more than just weaken the data set that is learned from, it can invalidate it by introducing our earlier-mentioned issue of bias.

Outliers can be some of the most troubling issues because of the subtlety of inclusion and the disastrous effects of incorrect removal. Often with government data use, the infrequent events can be the ones of most importance in keep. Examples are predicting trends or spikes in global exchange rates and detecting fraudulent or criminal activity. Outliers are left toward the end of this chapter and not covered in as much detail, not because of their unimportance, but because of the familiarity and experiences required to identify and handle them correctly. As an analyst starting out, the intuition and familiarity must be built with good data such that the techniques and correct applications are known. Once the tooling and mythology is familiar, one can move on to address problems that exist with the inputs and data. Attempting to address concerns with the data while also learning the techniques would leave a sense of ambiguity in which one lacks the discriminatory power to know where the issue is arising from.

8. NONREPRODUCIBILITY AND STATISTICS HUNTING

Try anything enough number of times and you really might see it once, most test tolerances are set to have a few percent chance of being incorrect due to randomness. As trivial as it is to mention reproducing important results as necessary, skipping it is currently a common bad habit. This often happens from root issues such as budgets or time constraints, but having good rote with reproducing data can protect important decisions where the consequences of being wrong are too high. There is also the ethical issue of where this happens with intent, where one latches on to the only single data point that tells the intended story. In the scientific community, the peer review system attempts to deal with this, but unfortunately, the government lacks a similar framework where decisions are always validated by others being implementation.

9. CONCLUSION

Machine learning has progressed readily in recent years, picking up the pace from its early conception as a result of the increasing data sets and powerful hardware. "I think AI is akin to building a rocket ship. You need a huge engine and a lot of fuel. If you have a large engine and a tiny amount of fuel, you won't make it to orbit. If you have a tiny engine and a ton of fuel, you can't even lift off. To build a rocket you need a huge engine and a lot of fuel. The analogy to deep learning is that the rocket engine is the deep learning models and the fuel is the huge amounts of data we can feed to these algorithms" (Ng, 2015).

In closing, it is reiterated that the government is entering an era where the abilities of a single analyst with appropriate knowledge and tooling can outperform what teams used to accomplish. "AI is the new electricity. Just as 100 years ago electricity transformed industry after industry, AI will now do the same" (Ng, 2016). Despite great tooling, the methodology is still not foolproof and mistakes will be made during the overhaul of processes and decision making. This chapter has surveyed some of these statistical challenges to help build an initial concept from which further investigation can be fostered.

REFERENCES

Andrew Ng Interview, 2015. Available from: https://www.wired.com/brandlab/2015/05/andrew-ng-deep-learning-mandate-humans-not-just-machines/.

Data Governance Framework, 2017. Available from: http://www.datagovernance.com/.

Encyclopedia of Mathematics, 2017. The European Mathematical Society. Springer. Available from: https://www.encyclopediaofmath.org/.

Fisher, R.A., 1936. The use of multiple measurements in taxonomic problems. Annals of Eugenics 466–467.

Keynes, J.M., 1939. The General Theory of Employment, Interest and Money. King's College, Cambridge.

Liebowitz, S., Margolis, S., 1999. Winners, Losers, and Microsoft: How Technology Markets Choose Products. Book published by the Independent Institute. ISBN: 978-0945999843.

Ng, A., 2016. Quote from Speech at Stanford MSx Future Forum. Chief Scientist at Baidu, Coursera co-founder, and Stanford Professor.

Orange, 2017. Version 3.3. Available from: http://orange.biolab.si/.

Samuel, A.L., 1959. Some studies in machine learning using the game of checkers. IBM Journal of Research and Development 3, 210–229.

The NSA, 2017. Available from: http://web.archive.org/web/20170310132753/https://nsa.gov1.info/utah-data-center/.

United States Government Accountability Office, 2016. GAO Cost Estimating as Assessment Guide. GAO-09-3SP.

USA 2010 Census Timeline, 2010. Available from: https://www.census.gov/2010census/partners/pdf/OperationalTimeline.pdf.

USA Jobs, 2017. Open Positions Requiring Excel Experience. Available from: https://www.usajobs.gov/Search/AdvancedSearch/.

Chapter 4

Making the Case for Artificial Intelligence at Government: Guidelines to Transforming Federal Software Systems

Feras A. Batarseh, Ruixin Yang
George Mason University, Fairfax, VA, United States

Intelligence is the Ability to Adapt to Change

Stephen Hawking

1. MOTIVATIONS AND OBJECTIONS

When the words Tesla or Edison is uttered, we think electricity; when we hear Newton, we think physics; similarly, when we hear Turing, we think Artificial Intelligence (AI). In his famous manuscript "Computing Machinery and Intelligence" (Turing, 1950), Turing paved the way for many scientists to think about AI. In his paper, Turing asked a question that is still under worthy amounts of debate: "Can Machines think?" As Turing navigated through different parts of pros and cons of proposing such a question, and as he claimed it as a noble cause, he warned against such an endeavor. Turing presented nine major challenges to AI, which are still relevant. These challenges included arguments from the informality of behavior, consciousness, mathematical challenges, theological, and other possible arguments that he anticipated would hinder the progress of AI. The most interesting one (and relevant to this chapter) was what he referred to as "Lady Lovelace's Objection" (Turing, 1950). Lady Lovelace's objection was that machines "can never do anything really new." In his paper, Turing goes into arguing for and against that perception, claiming that who knows if there is ever anything "new under the sun" to begin with. He challenges Lovelace's claims to define what "new" is, given that most of what we do is a follow-up of what someone else did already. What is very thought provoking, though, is

Federal Data Science. http://dx.doi.org/10.1016/B978-0-12-812443-7.00004-1

that Turing was arguing against what Lovelace referred to as the "Analytical Engine" (originally created by Charles Babbage) that she studied in 1842. Fast forward to this day, the promise of AI that was originally proposed by Turing is today transforming into a new age of Analytical Engines, Big Data, and Data Science.

As most industries are jumping on the wagon of AI, moving toward AI is becoming an inevitable reality. However, where is the federal government from all of that? Very few signs of federal AI projects are found. Before observing what the government is doing with AI, let us define AI. Based on the adopted definition of the Artificial General Intelligence Conference, "Intelligence is the ability to achieve complex tasks in complex environments" (Goebel et al., 2016). On the other hand, when the word "intelligence" is uttered anywhere inside the beltway in Washington, DC, the speaker must be referring to the intelligence community: a group of 16 separate government agencies (such as the Central Intelligence Agency, the CIA). To that point, it seems that an effort to injecting AI into governmental systems is a task that needs to start from scratch.

As more software tools become incrementally available to the government; as more data are created by the American public in every way, shape, and form; as the need to mine such data and form intelligent decisions based on them increases; and as many accountability, privacy, management, and security issues exacerbate at the government, using AI methods at the federal level is now an undeniable necessity that is already knocking on the doors. If the right data are in place, all types of AI algorithms could be applied, and that can unlock the government of the future, with better operations, better citizen service, and more efficient policy making. In a thesis published at Princeton University, data structuring was identified as one of the major challenges at federal agencies (Yu, 2012): "When government does collect and publish data in a reusable way, government enables third-party stakeholders like advocates, academics, journalists and others to powerfully adapt data in any way they see fit using the latest technologies, and to add value in unexpected ways. Third parties can use government data to experiment in parallel, in order to discover what innovations work best in changing technological environments. However, governments may be unwilling to publish structured datasets for various reasons, even if the data are already published in other public mediums, like on a web site or in print. In these cases, the traditional approach is to lobby the institution for changes in its publishing strategy. Alternatively, it may sometimes be possible to design and use software to create structured data from the outside, however painstaking the process." Fig. 4.1 shows where certain countries in the world are in terms of deploying data openness and developing projects based on data science (Kim et al., 2014). A study published in 2015 (Attard et al., 2015) showed that the following countries have the most "open" data: (1) the United Kingdom, (2) France, (3) New Zealand, (4) Norway, (5) Germany, and (6) Taiwan. The United States is evidently still falling behind.

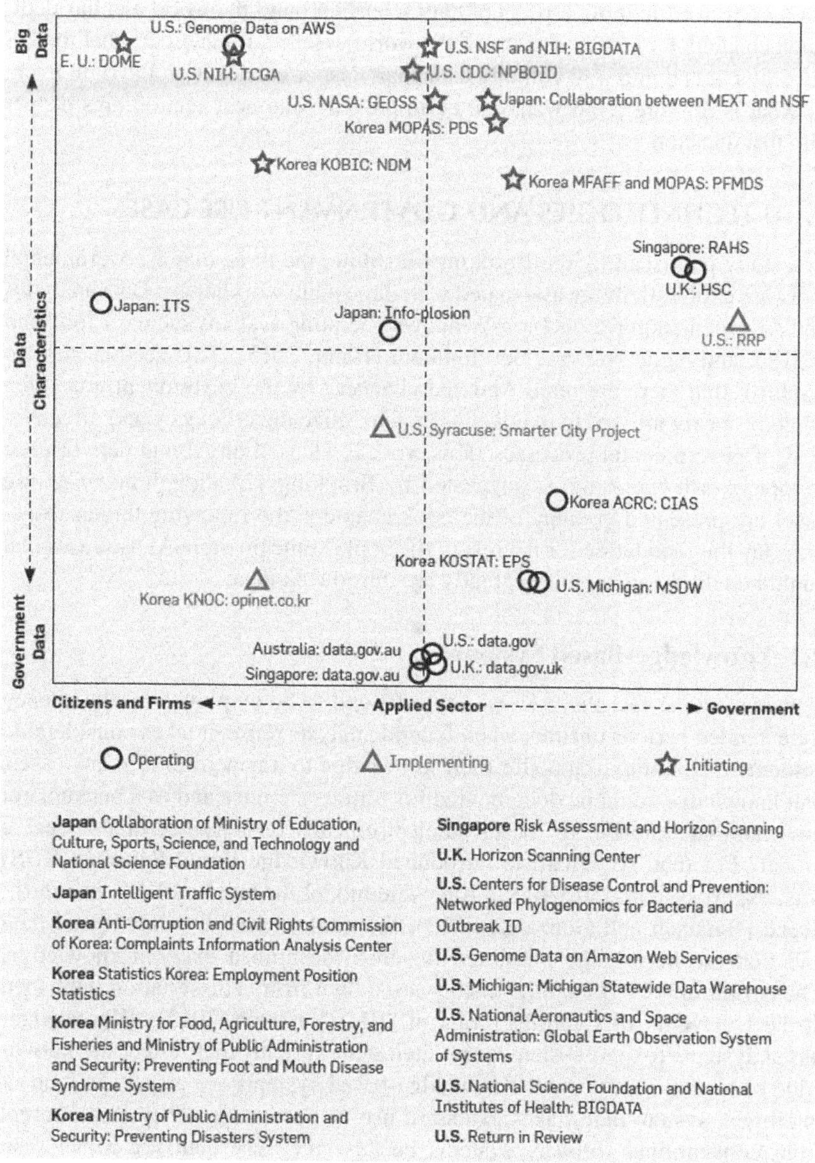

FIGURE 4.1 Data science at world governments (Kim et al., 2014).

In addition, based on Fig. 4.1, the United States has only one major (*government-wide*) data science project that is operating completely, which is the public website *www.data.gov*. Other countries have similar operating projects, whereas the United States has many projects that are in the initiating stage, so the federal government might be catching up with other world governments (this book goes over multiple aspects of data science around the world and has dedicated chapters for certain regions of the world). Hence, it may seem that the US government is doing well with big data analytics, or at least is on the right track, so what is missing? And what can be improved? The next section digs deeper into that question.

2. AI TECHNOLOGIES AND GOVERNMENT USE CASES

In a study published by the Brookings Institute, the three major governmental software explorations are associated with three main workhorses: Data analytics and AI, Open-source tools (non-Windows operating systems such as Linux and UNIX), and Agile software development (Hahn, 2015). This section aims to establish that there are many "missed chances" by the government, ones that could be easily utilized to exponentially maximize the efficiency and effectiveness of governmental processes. This work is focused on AI and data science (a topic worth exploring, as suggested by Brookings). Although many AI use cases are presented in many of the book chapters, the following three subsections lay the foundation for the rest of the book; some possible AI use cases that could benefit the government greatly are introduced next.

2.1 Knowledge-Based Systems

Considering the fact that federal analysts tend to be employed by the agency for extended periods of time, when federal analysts retire, it takes considerable amounts of training to transfer their knowledge to a new hire. In many cases, that knowledge could be documented for future reference and as a backup, but how can that "knowledge" be saved electronically or presented in a shareable model? For that, AI scientists introduced Knowledge-Based Systems (KBS) (i.e., expert systems). In 2013, a life cycle model for building KBS was introduced (Batarseh and Gonzalez, 2013); the model ensured validation of data and that the knowledge presented matches the human expert's knowledge. The definition of KBS and its history are drawn from a dissertation published at the University of Central Florida in 2011 (Batarseh, 2012): "Knowledge-based systems (expert systems) are intelligent systems that reflect the knowledge of a proficient person. Knowledge-based systems are a specific kind of intelligent system that makes extensive use of knowledge. They are different from conventional software systems because they use heuristic rather than algorithmic approaches for decision making. Furthermore, knowledge-based systems separate the knowledge from how it is used. The idea of a general problem solver (GPS) was introduced during the early 1960s. This idea used generic search techniques aided by heuristic knowledge to solve problems. DENDRAL, developed during the early 1970s, realized the internal structure of unknown compounds. The GPS and the DENDRAL experiences were instrumental in the development of MYCIN, a system that diagnosed blood disorders. MYCIN is a landmark medical rule-based system developed at

Stanford University. More importantly, MYCIN influenced the creation of the field of knowledge-based systems. Using the MYCIN experience, a number of knowledge-based systems in other domains were developed during the 1970s and 1980s such as HEARSAY (for language understanding), PROSPECTOR (for geology), XCON (for systems configuration) and GenAID (for electrical equipment diagnosis). A great number of expert systems have been developed since that time. The main component of the knowledge-based system is the knowledge base. Often, the knowledge reflected is that of an expert. KBS consist of an inference engine and a knowledge base. Inference engines act as the main controller of the system, where the knowledge is manipulated to address specific problems or answer specific questions. The inference engine contains the problem solving knowledge. The knowledge base is typically elicited from an expert or from an otherwise knowledgeable person in a certain domain. The knowledge is elicited, represented and built by the knowledge engineer, in a process known as knowledge engineering. Knowledge engineers can be said to be computer systems experts whom are responsible for representing knowledge in a computer system in order to solve problems that require human expertise" (Batarseh, 2012).

KBS could be used at many places at the government, for instance, *to standardize data at a federal agency*. Most federal agencies use data from multiple other agencies. Data are usually shared between databases, and it takes major efforts in some cases to move big amounts of data around. There are multiple issues concerned with that:

1. Data consistency among federal agencies: different agencies have different copies of the same data, how does an agency ensure consistency? Also, how does an agency obtain only what is needed by its analysts and not all the data from the source?
2. Data validity: when data move, data change! How do federal analysts know that the data are valid? How can they ensure that the numbers and nouns are up to date?
3. Data description and standards: different agencies have different data formats, time formats, and names for certain entities; in addition, different agencies care about different aggregation levels of the data. How does an agency make sure that the data are formatted and normalized according to its needs?

This section of the book addresses most of those three use cases and *answers* them in great detail. In some parts, it focuses on agricultural sciences applications at the US Department of Agriculture (as a suggested suitable starting point for the government).

2.2 Big Data

Another major technology that is apparently key for the US government is big data. The government seems to be trying to invest in this field; in a report published in 2016 (Patil, 2016), the office of the president shared the anticipated common challenges of injecting big data science at the government. The report (led by DJ Patil, the first US Chief Data Scientist, hired in 2015) highlighted the following challenges:

1. Sharing data to foster collaboration
2. Trusting data to drive decisions
3. Data resources to advance data-driven culture
4. Access to data tooling for increasing capabilities

These four challenges are addressed in this book in an explicit manner. The first challenge is addressed through an automated life cycle for streaming, which is introduced later in this book; in addition, data trust and resources are ensured through validation methods, also presented at many instances in the book. Access, tools, and all types of technologies are correspondingly discussed and evaluated across the book. However, before suggestions are provided, it is important to understand what tools the government is currently using for big data. Besides rational databases, SQL, Oracle tools, and the traditional Database Management Systems, is the government using the de facto of big data—Hadoop? (Hadoop Packages, 2016; Map Reduce, 2016) No strong evidence of that was found.

Fig. 4.2 illustrates Hadoop use cases for the US government, such as National Security and Defense, Federal Aviation Administration, and the Judicial System, and explores how and where Hadoop can be used. The famous four V's of big data (Velocity, Variety, Volume, and Veracity) form the opportunity and the challenge of applying this technology. Owing to the size and unstructured and nonrepetitive nature of the data, Hadoop could be very helpful; Swish Data (among others such as Map Reduce and Horton Works) [19, 20 and 21] published an article identifying multiple use cases for the government using Hadoop, some of which are illustrated in Fig. 4.2. Besides the ability to organize enormous amounts of data, Hadoop also handles unstructured data (which is a major characteristic of big data). However, because of the inability to use open source tools such as Hadoop, software engineers and researchers at federal agencies have to stick with commercial alternatives. Hadoop has the ability to execute advanced AI algorithms through packages such as Spark, which is a comprehensive AI and machine learning set of algorithms (i.e., a library called MLlib) that couples well with Hadoop.

2.3 Machine Learning and Data Analytics

As it is evident from the title, advanced analytics methods constitute the majority of this book. Big data is usually coupled with data analytics, and in some cases both terms are wrongly used interchangeably. In one paper (Batarseh et al., 2017), a comprehensive model for building government systems through data analytics was introduced (FedDMV: Federal Data Management

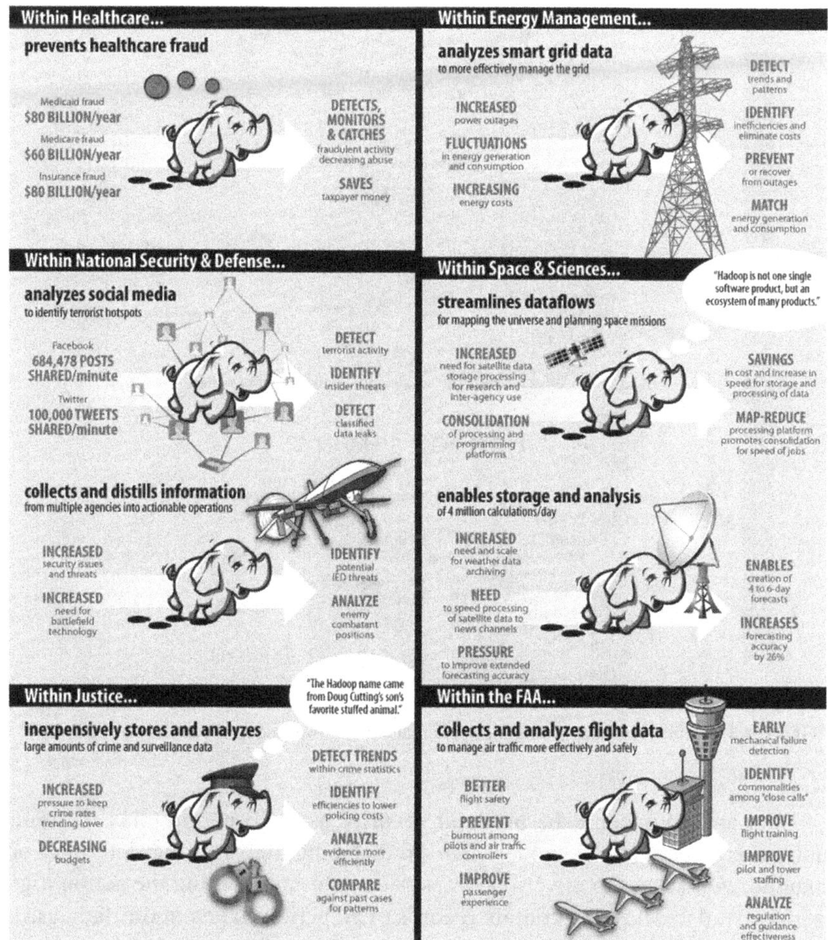

FIGURE 4.2 Hadoop use cases at the government (Swish Data, 2015).

and Validation), shown in Fig. 4.3. The main idea of the model is to use data analytical algorithms to improve federal data management and information validity. In Fig. 4.3, ADT stands for Analytics Driven Testing.

Another example was published in 2015, which demonstrated how data analytics can be used to evaluate health care policies at the state level (Batarseh and Latif, 2016). The paper used regression methods to build models that predict visits to public hospitals and a correlation model to view the effect of certain policies on public health. The results of the paper indicated that certain states got affected by the Affordable Healthcare Act more than others. Furthermore, it showed that certain states are in a critical need for more Medicaid and Medicare funding (such as South Carolina and Maine), versus ones that are more stable in that aspect (such as Texas and Washington) (Batarseh and Latif, 2016).

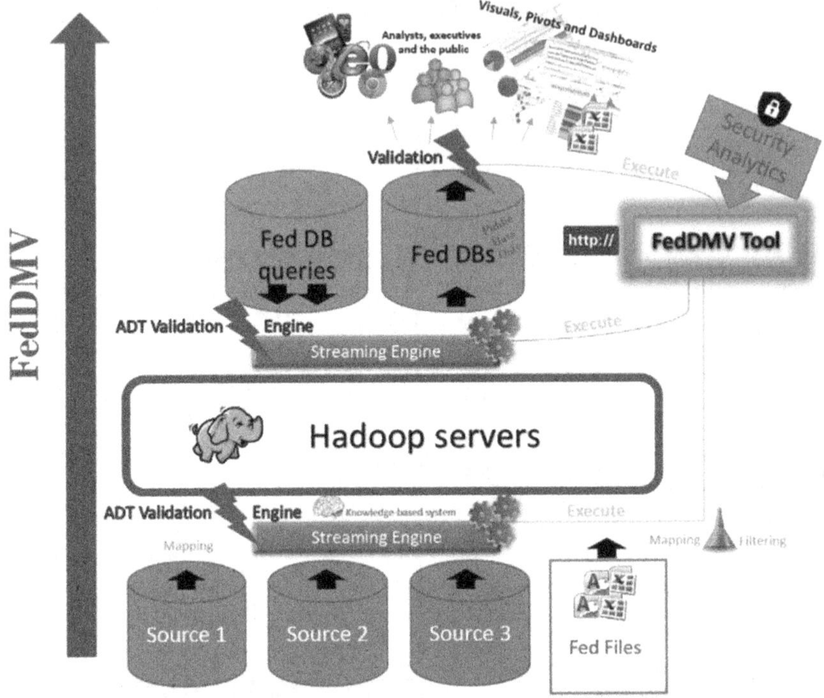

FIGURE 4.3 Example of a data analytics potential at the government (Batarseh et al., 2017).

Data analytics could be used for security as well; certain classification and pattern recognition models can predict potential hacking activities or unauthorized access to governmental systems through mining the action logs (a group of data files that contain records of all actions taken in the database). There are many other cases where data analytics could be used in improving the quality of federal systems; this book introduces multiple other examples throughout.

3. CONCLUSIONS

Although it is difficult to tell how many AI projects currently exist at the government, in a report published by the White House in 2012 (USDA and The White House, 2016), 70 projects were reported at different agencies to have an implementation of some sort that includes big data and data analytics. For example, there are many visualizations available at government websites that can help the users get information; nonetheless, what is proposed in this book is not a simple visualization or a small data calculation, rather, an A-to-Z set of advanced analytics that could completely transform policy making at the federal and state levels of government.

One would assume that there are many confidential or top secret projects that utilize forms of AI, especially in defense, army, navy, marines, and air force. Such projects might not be reported; but regardless of that, there is still a wide array of existing legacy systems that could be updated, and many other future ones that could be developed with AI in mind. Although this chapter suggested only three areas of AI for the government, Knowledge-Based Systems, Big Data, and Advanced Analytics (Machine Learning), the book touches on many other aspects of AI, and points to ways in which it could be used at the government. Adopted from a report published in i360Gov.com (I-360 Government Reports, 2016), Fig. 4.4 shows the main benefits of using big data analytics at the government (an increase of at least 20%, and up to 59% in all aspects of processes improvement). However, to accomplish that, data need to be available, structured, and in place (referred to as "linked data" in Fig. 4.5). The majority of federal data, however, are not readily available. Although many agencies claim to have data available, it is usually not ready for processing or analysis.

When users navigate through *data.gov*, or any other government data outlet, they would find that most files are in PDF format, MS word, and other formats that cannot be used for analysis (such as XML or CSV). Hence, providing data in the right format should be a priority for federal agencies going forward. That transformation would enable data openness, and increase the public's trust and overall accountability. For data to be used in AI, Machine Learning, and Data Analytics, they need to be *Structured*, *Open*, and *Linked*. Fig. 4.5 shows how only a small percentage of governmental data match that requirement

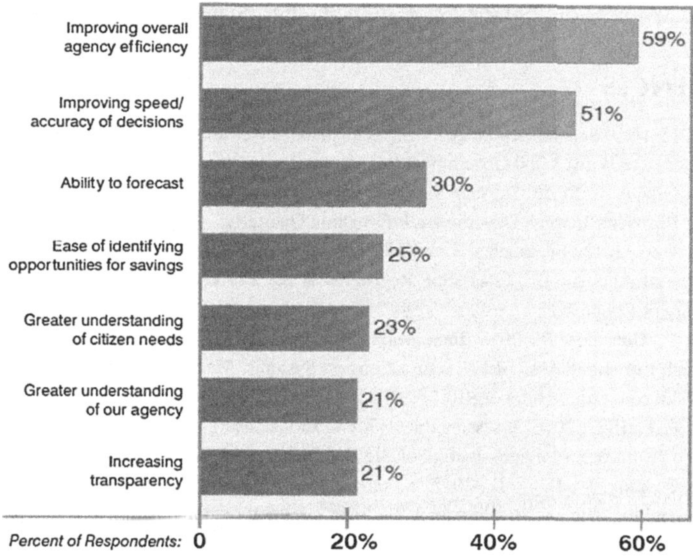

FIGURE 4.4 Predicted improvement percentages at government after using data science (I-360 Government Reports, 2016).

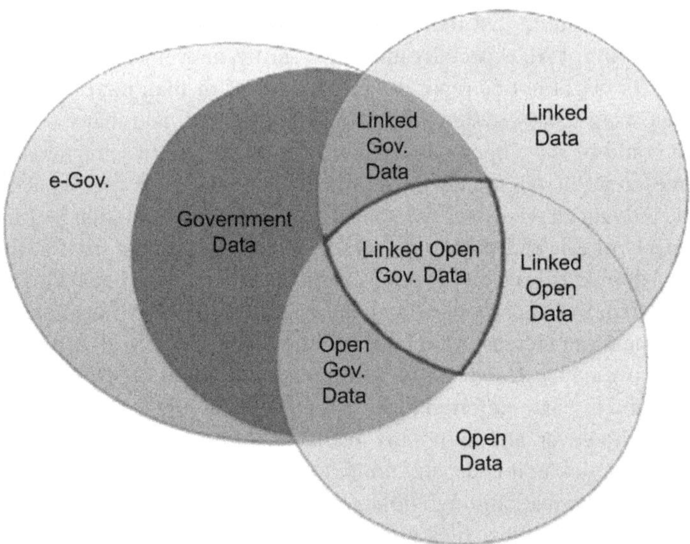

FIGURE 4.5 Current federal data: lots of smoke but no fire (Attard, 2015).

(the small highlighted area in the middle of the Venn diagram). Considering all these challenges and the recommendations given in this chapter, this book goes over detailed methods and guidelines to inject more data science and AI into governmental systems to improve the overall accountability, openness, security, and effectiveness of policy making in the United States.

REFERENCES

Attard, 2015. I-360 Special Report, 2013. Big Data in the Government Sector: Big Opportunity, Big Challenge or Both? I-360 Government Reports.

Attard, J., Orlandi, F., Scerri, S., Auer, S., 2015. A systematic review of open government data initiatives. Elsevier's Journal, Government Information Quarterly.

Batarseh, F.A., 2012. Incremental Lifecycle Validation of Knowledge-Based Systems through CommonKADS (Ph.D. Dissertation Registered at the University of Central Florida and the Library of Congress).

Batarseh, F., Gonzalez, A., 2013. Incremental lifecycle validation of knowledge-based systems through CommonKADS. IEEE Transactions on Systems, Man, and Cybernetics: Systems 43 (3), 643–654. http://dx.doi.org/10.1109/TSMCA.2012.2211348.

Batarseh, F., Latif, E., 2016. Assessing the quality of service using big data analytics: with application to healthcare. Elsevier's Journal of Big Data Research 4, 13–24.

Batarseh, F., Yang, R., Deng, L., 2017. A comprehensive model for management and validation of federal big data analytical systems. Springer's Journal of Big Data Analytics. http://dx.doi.org/10.1186/s41044-016-0017.

Goebel, R., Tanaka, Y., Wolfgang, W., 2016. Lecture notes in artificial intelligence series. In: The Ninth Conference on Artificial General Intelligence, New York, NY.

Hadoop Packages, 2016. Available from: http://hadoop.apache.org/.

Hahn, R., 2015. Government Policy toward Open Source Software: An Overview. Brookings Institute (Chapter 1).

I-360 Government Reports, 2016. Available from: http://www.i360gov.com.

Kim, H., Trimi, S., Chung, J., 2014. Big data applications in the government sector. Communications of the ACM 57 (3). http://dx.doi.org/10.1145/2500873.

MapR, 2016. Big Data and Apache Hadoop for Government. Available from: https://www.mapr.com/solutions/industry/big-data-and-apache-hadoop-government.

Patil, D., 2016. Advancing Federal Data Science Best Practices: A Report on Common Challenges, Product of the Committee on Technology of the National Technology and Science Council. The White House.

Swish Data – Data Performance Architects, 2015. Hadoop Uses Cases: Big Data for the Government. Available from: http://www.swishdata.com/index.php/blog/article/hadoop-use-cases-big-data-for-the-government.

Turing, A.M., 1950. Computing machinery and intelligence. Journal of the Mind 59, 433–460.

United States Department of Agriculture and The White House, 2012. Big Data across the Government. Executive Office of the President.

Yu, H.M.T., 2012. Designing Software to Shape Open Government Policy (A Dissertation Presented to the Faculty of Princeton University in Candidacy for the Degree of Doctor of Philosophy. The Department of Computer Science, Princeton University).

Section 2

Governmental Data Science Solutions Around the World

Chapter 5

Agricultural Data Analytics for Environmental Monitoring in Canada

Ted Huffman[1], Morten Olesen[1], Melodie Green[1], Don Leckie[2], Jiangui Liu[1], Jiali Shang[1]
[1]*Agriculture and Agri-Food Canada, Ottawa, ON, Canada;* [2]*Natural Resources Canada, Victoria, BC, Canada*

To reap the benefits of Big Data, it's important to ensure this is publicly available and shared with research and development partners. Only then will we be able to create a rich data ecosystem to support the knowledge-intensive and location-specific enterprise of agriculture.

Bill Gates

1. INTRODUCTION

As a federal government department cooperating with provinces under an agricultural policy framework, Agriculture and Agri-Food Canada (AAFC) has a variety of program and policy needs related to environmental assessments. The development of the science and data to support the framework, as well as to address international commitments such as national greenhouse gas inventory reports to the United Nations Framework Convention on Climate Change and agri-environmental reports to the Food and Agriculture Organization, generally falls within AAFC's Sustainability Metrics program. To fulfill the various needs, national soils, climate, hydrographic, topographic, land use, and activity data such as farming practices are required at various spatial scales. The data need to be temporal, spatially compatible, and accessible for use in geospatial process modeling and statistical analyses to provide estimates and predictions of the impact of agricultural production on air, water, and soil quality.

A national ecostratification framework has been established (AAFC, 2013a) as a hierarchical classification of the physical environment to serve as the base

Federal Data Science. http://dx.doi.org/10.1016/B978-0-12-812443-7.00005-3

for modeling at different levels of detail. Other data layers such as long-term climate normals, hydrography, general land cover, and growing season indices are tied to this framework. Other socioeconomic or "activity" data, such as crop areas, crop rotations, tillage practices, livestock numbers, and fertilizer and manure applications, have traditionally been obtained through the Census of Agriculture (Statistics Canada, 2016), but these data have limitations in that they are available only every 5 years and are on a different spatial framework than the ecostratification. To interpolate the census data to the ecobase with any degree of reliability, we recognized that we needed a series of high-resolution, high-accuracy land use maps to serve as a filter between the census framework, which is based on the cadastral survey and the ecoframework, which is based on natural landscape features.

With the advent of the Big Data era, particularly with respect to the rapid development of a wide variety of earth observation platforms, sensors, and interpretation algorithms, has come a plethora of land cover, land use, vegetation type and biomass maps. In Canada, a variety of land cover/land use products have been compiled at different times, for different extents, and for different purposes. One of the earliest products is the visually interpreted aerial photography-based 1966 Present Land Use series of the Canada Land Inventory (Natural Resources Canada, 2014), whereas a variety of different products have been generated more recently from satellite imagery. These include the current series of annual crop maps produced by AAFC (2013b), as well as a variety of products of different resolutions, classification schemes, areal coverages, and date stamps (e.g., Ontario Ministry of Natural Resources, 2008; Wulder et al., 2008; AAFC, 2009; Pouliot et al., 2014; Natural Resources Canada, 2015a).

Despite the number of Land Cover/Land Use (LC/LU) map products available, their application in temporal studies such as national greenhouse gases inventory reports or climate change mitigation and adaptation studies is fraught with difficulty. In the first place, the different spatial scales and classification schemes (especially with respect to land cover vs. land use) render them as barely comparable. For example, a "Herb" class of one map may correspond to both the "Cropland" and "Grassland" classes of another, or a map of 30-m resolution may identify small patches of "Forest" within a "Grassland" pixel of a 250-m-resolution map. Even if the classification schemes can be aligned, estimates of class areas often differ widely, especially for small areas, because the accuracy of production-level satellite-based maps seems to be currently maximized in the range of 75%–85% (see, for example, AAFC, 2009; Friedl et al., 2010; Kleeschulte and Bütt, 2006; Mayaux et al., 2006; Wickham et al., 2013). Under these conditions, monitoring land use over time, and especially determining land use change, from a comparison of maps provides unreliable results. The problem of "error accumulation" becomes even more acute when more than two maps are compared.

In an attempt to develop a stable land use platform on which to base environmental sustainability initiatives, our objective in this study was to compile as many relevant digital maps and information sources as possible, georegister all products on a 30-m base, and prepare a set of rules to be applied at each pixel to generate a "most probable" output class. We selected the six general land use classes of the Intergovernmental Panel on Climate Change (IPCC) (Forest Land, Cropland, Grassland, Wetlands, Settlements, and Otherland) (IPCC, 2006) with the additional separation of wetlands and water. The choice of land use categories as the map output meant that the interpretation of the more common input land cover classes such as "alvar", "herbs," or "shrubland," for which there is no clear use, relied particularly on other inputs. Table 5.1 provides an outline of Canadian definitions and thresholds of IPCC classes. Our target overall accuracy was 90%, with the goal of developing three output maps: 1990, 2000, and 2010.

TABLE 5.1 **Canadian Definitions and Thresholds for IPCC Land Use Categories**

Category	Definition
Forest Land	A minimum land area of 1 ha with a minimum width of 20 m, with woody vegetation with a minimum crown cover of 25% and a minimum height of 5 m. Forests also include systems with a vegetation structure that currently falls below, but could reach the threshold values (such as regrowth in harvested areas)
Cropland	Cropped land, including orchards, vineyards and agroforestry systems where the vegetation falls below the thresholds of Forest Land
Grassland	Rangeland and pasture land that are not considered Cropland, including systems with woody vegetation that falls below the threshold of Forest Land. Subdivided into Grassland Managed (grazing land) and Grassland Unmanaged (alpine meadows, tundra)
Wetland	Areas covered or saturated by water for all or part of the year. Subdivided into Wetlands (swamps, marshes, bogs, peatlands) and Water (permanent open water, including reservoirs)
Settlement	All developed or "built-up" land, including transportation and recreational infrastructure
Otherland	Bare soil, rock, ice, and all land areas that do not fall into any of the other five categories

IPCC, Intergovernmental Panel on Climate Change.

2. MATERIALS AND METHODS

The approach followed in our study included identifying and collecting suitable input maps and data, harmonizing the variety of legends to the extent possible, assessing class accuracies among the primary input source maps, registering all input products to a common base, developing rule sets and procedures designed to generate output maps on the basis of all input data, and assessing the accuracy of the output maps. The following sections provide details of each step in the methodology.

2.1 Input Data

A variety of LC or LU map products and data for the portion of Canada south of 60°N were identified as potential contributors to an integrated output product, including (1) GeoCover 1990, a commercial product of 30-m resolution produced by MDA Federal from Landsat imagery (University of Maryland, 1997–2016) and purchased for use in National Inventory Reports by Environment Canada; (2) Earth Observation for Sustainable Development of Forests (EOSD) maps produced by the Canadian Forest Service (CFS) (Wulder et al., 2008); (3) CanVec forest, roads, water, built-up and wetlands vector, and point layers (Natural Resources Canada, 2015b); (4) Land Cover for Agricultural Regions of Canada, circa 2000 30-m map produced by AAFC (AAFC, 2009); (5) Southern Ontario Land Resource Information System (SOLRIS) Land Use Data (Ontario Ministry of Natural Resources, 2008); (6) Annual Crop Inventory maps of 2010, 2011, and 2012 at 30-m resolution for agricultural regions of the country (AAFC, 2013b); (7) Land Cover of Canada 2005, a 250-m-resolution map of southern Canada (Pouliot et al., 2014) used only where no other inputs existed; and (8) Census of Agriculture (Statistics Canada, 2014). Considerable effort was put into identifying a suitable slope map, as we felt that such information could help make classification decisions in cases such as wetlands and croplands, but no data of sufficient resolution and accuracy were identified.

The project team consisted of domain experts from AAFC, Natural Resources Canada, Environment Canada and Statistics Canada, each with knowledge and experience in working with raster and vector maps and Earth Observation data. Two pilot sites (Outaouais, Quebec, and Meadow Lake, Saskatchewan) of approximately 1000 km^2 each were established to provide an independent indication of the accuracy of the primary input sources, to develop and test procedures and to identify technical issues involved in coregistering and interpreting the various products at the level of individual pixels. Based on the work in the pilot sites, a number of fundamental procedures were established: (1) prepare input maps by converting classes to the desired output classes as much as possible, (2) conduct accuracy assessments of input maps, (3) coregister all input products as carefully as possible, (4) develop a set of "rules" to define an output class from the specific and often conflicting input data at each pixel, (5) conduct contextual and discrepancy assessments and rectify conflicts, (6) compare output map class distributions with independent data and adjust as appropriate, and (7) conduct output map accuracy assessments. Each of these procedures is outlined in more detail in the following sections.

2.2 Input Preparation

Input preparation required the renaming of a variety of classes in different input maps to IPCC categories (Table 5.2), while retaining others (e.g., water, forest, settlements) as appropriate.

In most cases, conversion of input classes to output classes was straightforward and involved simply renaming. For example, Forest–Deciduous and Forest–Coniferous of GeoCover were both renamed Forest, as were the

TABLE 5.2 Examples of Different Input Classes and Their Renamed Class

Examples of Input Classes	IPCC Class
Forest–Deciduous; Forest–Coniferous; Broadleaf Dense Forest; Dense Deciduous Forest; Coniferous Dense Forest; Mixedwood Dense Forest; Broadleaf Open Forest; Coniferous Open Forest; Mixedwood Open Forest; Broadleaf Sparse Forest; Coniferous Sparse Forest; Mixedwood Sparse Forest; Coniferous Plantation	Forest
Agriculture–General; Pasture and Abandoned Fields; Annual Cropland; Perennial Cropland; Cultivated Agricultural Land; Orchards; Vineyards	Cropland
Rangeland; Native Pasture; Native Grass; Meadow; Alvar; Tallgrass Savannah	Grassland
Wetland–Permanent Herbaceous; Wetland–Herb; Wetland–Shrub; Wetland–Treed; Inland Marsh; Conifer Swamp; Deciduous Swamp; Open Fen; Bog; Peatland	Wetland
Urban/Built-up; Developed Land; Industrial Land; Residential; Roads; Railroads; Commercial Land; Buildings; Transportation, Recreational; Institutional; Airport; Extraction; Pits and Quarries	Settlement
Reservoir; River; Lake	Water
Exposed/Barren Land; Rock/Rubble; Beach; Permanent Ice; Bare Rock; Open Sand Barren and Dune; Non-Vegetated Land; Sparsely Vegetated Bedrock; Open Cliff and Talus; Bryoids; Open Shoreline	Otherland
Cloud; Shadow; No data	No data

IPCC, Intergovernmental Panel on Climate Change.

nine forest classes of EOSD. Similarly, all input classes relating to swamps, marshes, bogs, and wetlands were renamed Wetlands; all built-up, urban, developed, and settlement classes were renamed Settlements; classes depicting barren, exposed, rock, ice, or beaches were renamed Otherland, and agricultural classes such as cropland, pasture, cultivated, orchards, vineyards, and nurseries were renamed Cropland. The EOSD class "Herbs," which could be associated with settlement (golf courses, parks), grassland, or cropland, and the class "Shrubland," which could be forest, wetland, grassland, or otherland were retained as input classes to use the information content in rule development. Vector files from CanVec (water, wetlands, National Road Network) were rasterized at 30-m resolution and building point files were converted to appropriate-sized clusters of 30-m pixels and labeled as Settlement. Assessment of rasterized roads in the pilot sites indicated that rasterizing CanVec primary roads to three pixel widths, secondary roads to two pixels, and tertiary roads to one pixel provided the most accurate definition of road surface area.

2.3 Input Accuracy Assessment

Detailed accuracy assessment of the primary input maps at the Outaouais and Meadow Lake pilot sites was performed to supplement the documentation supplied with the different input products and to establish specific class accuracies where none were available. The confusion matrices were developed using randomly selected ground reference points visually interpreted from appropriate dates of aerial photos, Landsat imagery, and field survey. An example of the confusion matrices for GeoCover1990, EOSD2000, and AAFC LC2000 at the Outaouais site are shown in Table 5.3.

2.4 Coregistration

All map products were resampled to 30-m pixels using NAD 1983 datum, with the AAFC2000 product serving as the base layer. This allowed the generation of classes depicted on each input product at every pixel in the country. The National Road Network with a date stamp closest to the year of concern was "burned into" the primary input source for each of the three maps and the class was considered inviolate in further manipulations.

2.5 Rule Development

Rules were developed to be applied at each individual pixel, with the output class at each pixel being assigned a class based on the most accurate or most likely information for that pixel. Three "zones" in which rules varied slightly were identified; the agricultural zone, beyond which agriculture was known not to occur, the rangeland zone, within the agricultural zone and consisting of areas

TABLE 5.3 Confusion Matrices Showing the Accuracy of Each Class at the Outaouais Study Site for GeoCover1990, EOSD2000, and AAFC LC2000

		Ground-Truth							
		Forest	**Water**	**Cropland**	**Settlement**	**Wetland**	**Otherland**	**Total**	**User's Accuracy (%)**
GeoCover1990 (Overall Accuracy = 74.8%, Kappa Coefficient = 0.63)									
Map	Forest	572	4	42	13	52	0	683	83.7
	Water	0	84	0	0	2	0	86	97.7
	Cropland	32	1	358	13	14	3	421	85.0
	Settlement	1	0	0	87	0	0	88	98.9
	Grassland	3	0	14	2	4	0	23	0.0
	Wetland	3	1	3	2	17	1	27	63.0
	Otherland	0	0	2	1	0	0	3	0.0
	Shrubland	75	2	59	3	24	0	163	0.0
	Total	686	92	478	121	113	4	1494	
Producer's Accuracy (%)		83.4	91.3	74.9	71.9	15.0	0.0	74.8	

Continued

TABLE 5.3 Confusion Matrices Showing the Accuracy of Each Class at the Outaouais Study Site for GeoCover1990, EOSD2000, and AAFC LC2000—cont'd

		Ground-Truth							
		Forest	Water	Cropland	Settlement	Wetland	Otherland	Total	User's Accuracy (%)
EOSD2000 (Overall Accuracy=82.1%, Kappa Coefficient=0.73)									
Map	Forest	647	7	16	21	28	0	719	90.0
	Water	3	85	0	1	6	0	95	89.5
	Cropland	12	0	378	29	13	1	433	87.3
	Settlement	5	0	52	65	8	0	130	50.0
	Grassland	0	0	0	0	0	0	0	N/A
	Wetland	2	0	2	4	51	0	59	86.4
	Otherland	0	0	0	0	0	1	1	100.0
	Shrubland	13	0	26	11	6	1	57	0.0
	Total	682	92	474	131	112	3	1494	
Producer's Accuracy (%)		94.6	92.4	79.5	50.4	46.0	50.0	82.2	

AAFC LC2000 (Overall Accuracy = 84.7%, Kappa Coefficient = 0.78)

Map	Forest	581	2	3	10	15	0	611	95.1
	Water	21	90	4	4	16	0	135	66.7
	Cropland	4	0	437	0	0	0	441	99.1
	Settlement	7	0	2	102	0	1	112	91.1
	Grassland	5	0	20	0	0	0	25	0.0
	Wetland	14	0	0	1	54	0	69	78.3
	Otherland	22	0	0	5	0	1	28	3.6
	Shrubland	28	0	8	9	27	1	73	0.0
	Total	682	92	474	131	112	3	1494	
Producer's Accuracy (%)		85.2	97.8	92.2	77.9	48.2	33.3		

Cell values = n (ground-truth points).
Ground-truth data did not identify the map classes Grassland or Shrubland in the study site.

where soils data indicated that natural grassland could occur, and the nonagricultural zone. Several general principles were formulated: (1) a pixel class should not be changed unless evidence indicates, (2) preponderance (generally referred to as "vote") provides more evidence than accuracy (i.e., two sources of agreement at 85% accuracy overrules a different one of 87% accuracy), and (3) the rules should be based, in order of application, on (a) preponderance of evidence, (b) map accuracy, and (c) context.

We started rule development using only input sources for 1990 and 2000, with the intent of developing only those two output land use maps. Rule development began with an assessment of the distribution of input combinations. With all input maps coregistered, a spreadsheet detailing the number of pixels with each combination of inputs was prepared and sorted to identify the variety and distribution of input combinations. A draft rule for each combination based on preponderance of evidence and relative accuracies of the input classes was developed, applied in the pilot areas, and the results were presented to the interdisciplinary expert group. Some revisions to the "preponderance priority" rule were made based on input map accuracy, with those class/product cases showing greater than 90% classification accuracy (for example, all input maps for Water, EOSD, and SOLRIS for Forest, and AAFC LC2000 for Cropland) taking precedence over a predominance of the lower-accuracy Wetland and Otherland classes.

Rule development required that both differences in input classes as well as land use change be taken into consideration. Because we had only one input for 1990, we used 2000 information to supplement it by "backcasting," with the caveat of not eliminating legitimate land use change. For example, we considered the spontaneous appearance of Forest to be very unlikely, so pixels of Forest in 2000 (high accuracy) and Wetland, Cropland, or Otherland in 1990 (low accuracy) became Forest in 1990. Conversely, Settlement in 2000 did not necessarily indicate Settlement in 1990, as spontaneous appearance of Settlement is quite likely. LU changes such as Forest to Cropland, any class to Settlement, Grassland to Cropland, and Wetland to Forest were considered legitimate and retained, whereas changes such as Settlement to any other class, any other class to Grassland, and Otherland to Forest were considered not legitimate and eliminated. In practice, rules were formulated in an "If…then" format, as in "If 2 sources indicate forest and one indicates water, then the output should be forest," and were applied as an output designation for each combination of inputs in the spreadsheet.

The first set of rules related to the 1990 and 2000 input sources and generally consisted of preponderance and accuracy rules modified by the change rules; if a majority of inputs agree, then that class was assigned as output, and if all sources disagree, then the class with the highest accuracy was assigned as the output class. The zonation of the study area also affected the rules, as no cropland was allowed outside of the agricultural zone, all grassland within the "grassland zone" was designated as "Grassland-Managed," and all grassland outside of that zone was designated "Grassland-Unmanaged." In all cases, in an attempt to concur with IPCC guidelines by eliminating the shrubland class, a shrubland input was considered to support forest or wetland of another input, and in a direct discrepancy, any other class took precedence over shrubland. Similarly, "Herbs" supported Grassland, Cropland, or Settlement and "Wetland" supported water. An example of a spreadsheet with pixel count, input classes, and output classes as defined by the rules is presented in Table 5.4.

By the time of completion of draft 1990 and 2000 LU maps, national crop maps for 2010, 2011, and 2012 had become available and the decision was made to add a 2010 LU map. Evaluation of the crop maps using appropriate ground-truth points indicated that classification of the type of crop and grassland met the 85% stated accuracy of the maps, but the accuracy of mapping Settlement was somewhat lower and the identification of Forest, Wetland, Water, and Otherland was considerably lower. The crop maps were coregistered to the 1990 and 2000 output maps, and a similar set of rules following the same principles was developed to generate a 2010 LU map. In some cases of significant conflict between the two output maps and the crop maps, the original input sources for the 1990 and 2000 maps were consulted for resolution.

2.6 Contextual Assessment and Rectification

One of the most significant improvements in raster classification in the recent past has been the development of segmentation, or the recognition of clusters of similar pixels. This concept was employed in our study to eliminate small clusters of Cropland and Grassland Managed, as we considered that the appearance of very small fields of these classes within a larger block of a different class was unlikely and probably due to misclassification. Thus in the prairie region, where farm fields are generally larger than 60 ha, we identified independent patches of Grassland smaller than 10 ha (91 pixels) and Cropland smaller than 5 ha (46 pixels) and converted them to the dominant surrounding land use type. In the remainder of the country, where farm fields can be smaller, we eliminated patches of Grassland and Cropland of less than 2 and 1 ha, respectively.

Contextual modifications were also employed in the form of a majority filter to eliminate roads that had appeared across large water bodies as a result of ferry routes being classified as roads in the CanVec roads network. In a similar correction, all grassland within the grassland soils zone was renamed "Grassland Managed," whereas all grassland outside of the agricultural zone was renamed "Grassland Unmanaged." Any remaining Shrubland pixels (where all inputs indicated that class) and Herb pixels (where EOSD was the only input) were renamed using a majority filter.

The most difficult distinction was the correct classification of large blocks of harvested or burnt forest, which were generally designated as Cropland, Grassland, or Shrubland on input maps. Most of it occurred outside of the

TABLE 5.4 An Example of a Spreadsheet Showing Pixel Counts, Combinations of Four Inputs, and 1990 and 2000 Outputs as Defined by the Rules

	Inputs				Outputs	
Count (pixels)	**GeoCover1990**	**AAFC LC2000**	**EOSD2000**	**CANVEC (c.2000)**	**LU90**	**LU00**
89,320,256	Cropland	Cropland	Unclassified	Unclassified	Cropland	Cropland
73,097,451	Unclassified	Unclassified	Unclassified	Unclassified	Unclassified	Unclassified
72,871,204	Shrubland	Unclassified	Forest	Unclassified	Forest	Forest
44,420,304	Cropland	Cropland	Herbs	Unclassified	Cropland	Cropland
43,413,067	Forest	Unclassified	Forest	Unclassified	Forest	Forest
34,925,671	Water	Unclassified	Water	Water	Water	Water
15,696,365	Shrubland	Unclassified	Shrubland	Unclassified	Shrubland	Shrubland
12,269,881	Water	Water	Water	Water	Water	Water
12,215,505	Shrubland	Forest	Forest	Unclassified	Forest	Forest
8,651,350	Grassland	Grassland	Unclassified	Unclassified	Grassland	Grassland
8,264,079	Forest	Unclassified	Shrubland	Unclassified	Forest	Forest
8,088,053	Shrubland	Cropland	Unclassified	Unclassified	Cropland	Cropland

6,753,562	Grassland	Cropland	Unclassified	Unclassified	Cropland	Cropland
5,981,113	Shrubland	Unclassified	Wetland	Unclassified	Wetland	Wetland
4,345,932	Shrubland	Unclassified	Forest	Wetland	Wetland	Wetland
4,063,368	Water	Unclassified	Water	Unclassified	Water	Water
4,002,243	Shrubland	Cropland	Herbs	Unclassified	Cropland	Cropland
3,439,317	Shrubland	Grassland	Unclassified	Unclassified	Grassland	Grassland
2,509,282	Forest	Cropland	Herbs	Unclassified	Forest	Cropland
2,460,156	Water	Unclassified	Forest	Unclassified	Water	Water
2,297,535	Shrubland	Wetland	Wetland	Wetland	Wetland	Wetland
2,290,574	Forest	Shrubland	Forest	Unclassified	Forest	Forest
2,197,465	Cropland	Grassland	Unclassified	Unclassified	Cropland	Cropland
2,157,877	Shrubland	Shrubland	Forest	Unclassified	Forest	Forest
1,993,485	Forest	Forest	Unclassified	Unclassified	Forest	Forest
1,974,893	Wetland	Unclassified	Forest	Unclassified	Forest	Forest
1,973,660	Forest	Unclassified	Forest	Wetland	Forest	Forest
1,891,737	Forest	Forest	Forest	Wetland	Forest	Forest

agricultural zone, so conversion of the Cropland designation to Forest was straightforward, but grassland and shrub-covered wetland are legitimate land covers in the zone, and separating those from harvested forest was difficult. Visually, harvested areas were quite recognizable by the straight edges, but burnt areas were often indistinguishable from natural grassland or wetland. Overlaying a "Forest Fire" map (Natural Resources Canada, n.d.) provided some assistance, but it appeared that the mapped burned areas also encompassed natural grasslands and thus the area of forest was overestimated. Visual interpretation of anything other than very large blocks was impractical, and the problem persists.

2.7 Assessment of Class Distributions

With the completion of the three draft maps based on all spatial inputs, the distribution of each class was compared wherever possible with independent statistical sources at local, provincial, and national scales. The national Census of Agriculture (Statistics Canada, 2016), conducted as a self-administered questionnaire on every farm in Canada every 5 years, provides a comprehensive data set pertaining to agricultural activities, including the area of cropland and pasture, at various spatial scales. We identified the amount of cropland and pasture listed for each Census Division in 1991, 2001, and 2011 and compared it with the area mapped for the previous year to determine whether the land use distributions were aligned. It was not expected that the areas would match exactly, because the Earth Observation–based maps identified idle farmland, non-Census hobby farms, and open land in protected and First Nations communities as cropland or grassland. In most cases, areas mapped as cropland and grassland showed slightly higher estimates than the area of cropland and rangeland reported in the census, with a few exceptions. In areas along the Atlantic coast, the census identified hay and pasture in areas where the maps showed only Forest, Wetland, and Otherland. In those few areas we used imagery and aerial photography to visually identify and digitize open areas around residences as cropland. In other areas such as the mixed grassland/forest areas in the eastern foothills of Alberta and interior valleys of British Columbia, we found relatively large discrepancies between the area of mapped grassland and the area reported in the Census as unimproved pasture, with the maps showing a smaller area. We expanded the Grassland category into Forest by two pixels on all boundaries between the two, which improved but did not entirely eliminate the differences between the two estimates. We assume that the discrepancy relates to ranchers reporting pastured forest as unimproved pasture in the Census.

As a final step, we applied a 3×3 majority filter to all classes except Settlement and Wetland (assuming that these two classes could occur as very small areal units) over the entire study area to eliminate cases of single pixels of one class within a larger area of a different class.

2.8 Output Accuracy Assessment

After a final check to identify any intermap discrepancies, map accuracy at the pixel level was assessed using randomly selected ground-truth points of the appropriate year compiled from field survey or visual interpretation of aerial photography or high-resolution satellite imagery. For some classes such as Forest, Otherland, and Wetland, ground reference points for a later year were used for an earlier year as well, based on our assumption that these classes do not appear spontaneously. In the case of ground-truth points that occurred within a mixed pixel, we assigned two correct classes, rather than discarding them as "undefined." For example, a ground-truth point that fell within a mixed water/forest pixel was assigned two possible land uses: water or forest. The map pixel was considered correct if it depicted either water or forest, but was considered incorrect if it denoted any other class. We feel that this approach avoids creating a ground-truth bias toward more easily classified "pure" pixels, and thus provides a more realistic interpretation of map accuracy. Accuracy assessment of the LU1990 map was based on 7139 points, LU2000 on 7218 points, and LU2010 on 4064 points.

3. RESULTS AND DISCUSSION

The methodology outlined earlier resulted in three output land use maps: 1990, 2000, and 2010. Accuracy assessment of the resultant maps showed overall improvements over the input data, with improvements in some classes and lowered performance in others. The following sections provide details on the changes in accuracy, present an example of an output map, and discuss the potential of using the decadal maps for estimation of land use change.

3.1 Accuracy Assessment

The entire study area encompassed approximately 550 million hectares and 6 billion pixels. Coregistering and overlaying the 1990 and 2000 input sources identified approximately 4000 combinations of classes, with fewer than 100 combinations representing 90% of all pixels. Ten combinations were identical across all inputs and represented approximately 69% of the total area.

As expected, class accuracies reflected the accuracies of the input data, but the overall accuracy of the three output maps, at 84.0%, 87.1%, and 92.7% (Kappa=0.78, 0.82, 0.92) for 1990, 2000, and 2010, respectively (Table 5.5), surpassed the accuracies of all input maps (Table 5.3), thus showing the synergistic effects of integrating multiple data sources. In addition, the accuracy of the output maps was greater in each succeeding time period, showing the benefits of more sources of input data.

The highest output classification accuracies (>90%) from a producer's perspective were obtained for Water and Forest, which is consistent with the

TABLE 5.5 Confusion Matrices of the Three Land Use Output Maps

		Ground Truth								
		Forest	Water	Cropland	Settlement	Grassland	Wetland	Otherland	Total	User's Accuracy (%)
LU1990; Overall Accuracy = 84.0%, Kappa Coefficient = 0.78										
Map	Forest	2781	25	103	4	78	172	87	3250	85.6
	Water	32	1372	3	0	7	74	8	1496	91.7
	Cropland	56	6	524	18	26	40	5	675	77.6
	Settlement	23	2	37	190	1	9	8	270	70.4
	Grassland	12	3	35	0	177	9	13	249	71.1
	Wetland	113	9	26	5	29	859	30	1071	80.2
	Otherland	11	1	5	0	2	17	91	127	71.7
	Total	3028	1418	733	217	320	1180	242	7138	
Producer's Accuracy (%)		91.8	96.8	71.5	87.6	55.3	72.8	37.6		

LU2000; Overall Accuracy = 87.1%, Kappa Coefficient = 0.82										
Map	Forest	2870	18	84	3	61	109	60	3205	89.5
	Water	28	1181	27	0	5	55	6	1302	90.7
	Cropland	48	6	1004	15	27	35	3	1138	88.2
	Settlement	19	3	25	217	0	1	4	269	80.7
	Grassland	13	2	45	0	164	7	11	242	67.8
	Wetland	85	8	36	5	23	776	27	960	80.8
	Otherland	8	1	0	0	2	17	75	103	72.8
	Total	3071	1219	1221	240	282	1000	186	7219	
Producer's Accuracy (%)		93.5	96.9	82.2	90.4	58.2	77.6	40.3		

Continued

TABLE 5.5 Confusion Matrices of the Three Land Use Output Maps—cont'd

		Ground Truth								
		Forest	Water	Cropland	Settlement	Grassland	Wetland	Otherland	Total	User's Accuracy (%)
LU2010; Overall Accuracy = 92.7%, Kappa Coefficient = 0.92										
Map	Forest	1719	8	12	3	16	43	41	1842	93.3
	Water	12	839	3	0	0	14	3	871	96.3
	Cropland	20	1	429	0	18	2	0	470	91.3
	Settlement	5	0	5	50	0	0	0	60	83.3
	Grassland	8	0	7	0	82	5	4	106	77.4
	Wetland	24	3	3	2	1	464	3	500	92.8
	Otherland	7	1	1	2	2	19	182	214	85.0
	Total	1795	852	460	57	119	547	233	4063	
Producer's Accuracy (%)		95.8	98.5	93.3	87.7	68.9	84.8	78.1		

input data sources and reflects the high level of accuracy generally obtained for these classes in remote sensing applications. The lowest accuracies, in the 40%–60% range, were for Otherland and Grassland, although the accuracy for those classes improved to 60%–80% in LU2010. The relatively low accuracies in those classes parallel the poor results of the input maps and reflect the fact that, in addition to high-mountain snow and ice, Otherland commonly occurs in Canada as small patches of bare rock intermixed with forest and wetlands, whereas Grassland outside of the large rangeland tracts occurs as relatively small patches intermixed with forest, cropland, and wetland. The small size of Otherland and Grassland patches, and their mix with other classes, causes higher levels of confusion and this poor performance has prompted the initiation of several specific studies to improve grassland mapping in Canada.

Producer's accuracy for Cropland in the LU1990 map was only 72%, which reflects the poor accuracy of that class in GeoCover. Accuracy of Cropland in 1990 could possibly have been improved by "backcasting" cropland from 2000 to 2010, but we had rejected that approach because of the potential for missing change from forest and grassland to cropland. Accuracy improved to 82% in 2000 and 93% in 2010, reflecting the good results of the AAFC LC2000 and SOLRIS maps (not shown) and the influence of the 2010 crop maps. Producer's accuracy for Settlement remained at 85%–90% in all three output maps, which is an improvement over the 50%–80% of the input data and indicates that our rule stipulating "once Settlement, always Settlement" is essentially correct. However, identification of Settlement is still hampered by the fact that much of it occurs as small patches (individual homes, hamlets, industrial units, etc.) surrounded by cropland or forest. Error in the urban class is also the result of many large urban areas containing water bodies, open grass, and heavily treed areas within their boundaries.

Despite the limited extent of accuracy assessment of the input maps compared with the national assessment of the output maps, it is interesting and informative to review producer's versus user's accuracies. Generally, the output maps show higher producer's accuracies and equal or lower user's accuracies than the input maps, with the improvement overwhelming the losses to produce higher overall accuracies. For example, results for 1990 indicate that the integration procedure reduced commission errors in Forest, Grassland, Wetland, and Otherland, whereas it increased them in only Water, Cropland, and Settlement. Similarly, LU1990 reduced omission errors in Forest, Water, Settlement, Wetland, and Otherland and increased omission errors in only Cropland. The output map for 2000 showed similar results in reducing errors in several classes while increasing errors in fewer classes or to a lesser extent. The net result, as indicated by overall accuracies, was to create more balanced accuracies across all classes, even though in some cases (GeoCover1990 Cropland, EOSD2000 Forest, AAFC LC2000 Water) the input maps showed better performance than the corresponding output maps.

3.2 National-Scale Maps

The study resulted in three maps, one for each of 1990, 2000, and 2010, covering all areas of Canada south of 60°N. An example of an output map (2010), with provincial and ecozone boundaries, is presented in Fig. 5.1. Cropland occurs primarily in the Prairie Ecozone (10) and the Mixedwood Plains Ecozone (8), with extensions into the Boreal Plains Ecozone (9), the Atlantic Maritime Ecozone (7), and the southern parts of the Pacific Maritimes Ecozone (13). Forest is the major land use in most other ecozones, whereas Wetland is found primarily in the Boreal Shield (6), Taiga Plains (4), and Hudson Plains (15) ecozones. Grassland is found primarily in the Prairie Ecozone (10), Otherland in the Montane Cordillera (14) and Taiga Shield ecozones, and Settlement in the Mixedwood Plains and Atlantic (7) and Pacific (13) Maritime ecozones.

A summary of the total area in each of the seven land use classes for the three maps is presented in Table 5.6. Within the area mapped, which included both land and water, Forest shows the largest proportion (54.0%), followed by Water (23.7%). The proportion of area under other uses ranges from 7.2% for Cropland to 0.8% for Settlement. From 1990 to 2010, there was a noticeable increase in the area of Settlement (19.4%) and Cropland (2.4%) and a decrease in Grassland area (5.9%).

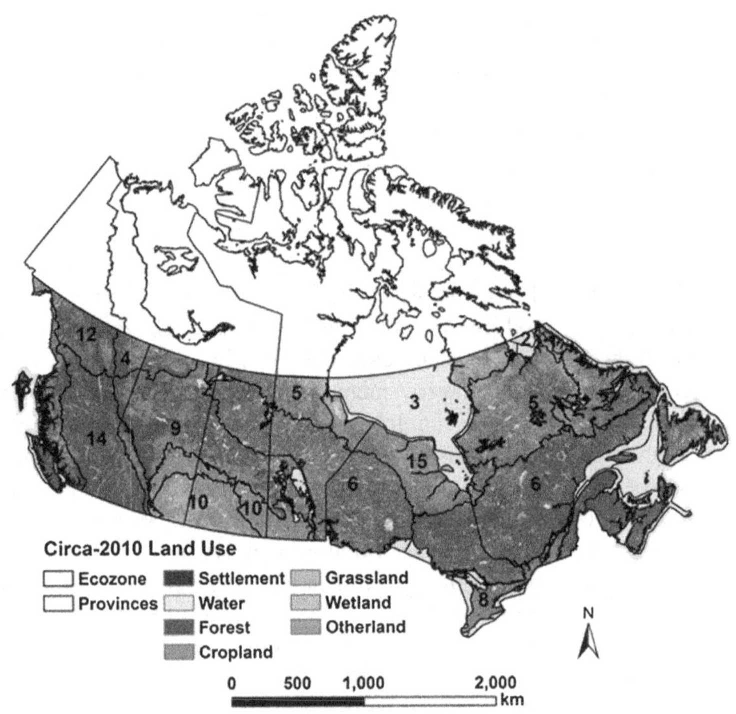

FIGURE 5.1 An example of an output land use map of Canada south of 60°N latitude, 2010.

TABLE 5.6 Class Areas (MHa) Mapped for Canada South of 60°N Latitude on the 1990, 2000, and 2010 Land Use Maps

Year	Forest	Water	Cropland	Settlement	Grassland	Wetland	Otherland
1990	369.508	162.053	48.060	4.982	13.665	47.423	37.206
2000	368.689	162.075	48.830	5.517	13.264	47.326	37.196
2010	368.426	162.147	49.210	5.947	12.859	47.124	37.182

3.3 Land Use Change

There is an interest, especially in the community of specialists working on national greenhouse gases and other environmental issues and policies, to use the three decadal land use maps to identify and quantify rates of land use change. Because some of the rules were expressly developed to eliminate "unrealistic" land use change such as the spontaneous appearance of Forest or the disappearance of Settlement and because accuracy of the maps is estimated to be near or over 90%, it appears that the maps might be used for land use change detection. Fig. 5.2 shows two examples of land use change identified by the maps between 1990 (left) and 2010 (right). The two upper maps (A) show land use change for the "Golden Horseshoe" around the western end of Lake Ontario in southern Ontario. This area encompasses Toronto and the nearby cities of Hamilton, Guelph, Kitchener-Waterloo, Cambridge,

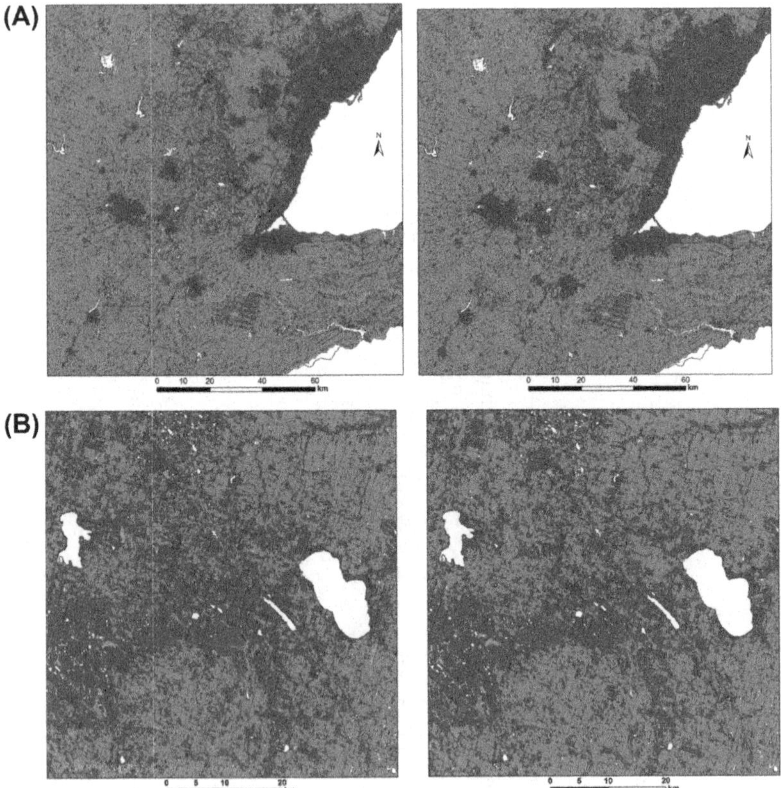

FIGURE 5.2 Land use and land use change depicted at example locations in Canada. (A) "Golden Horseshoe" at the western end of Lake Ontario and (B) a portion of Wetaskiwin County, Alberta. Left map is 1990, right is 2010.

and Brantford and is one of the most populated and urbanizing regions of Canada. The maps show a significant expansion of the Settlement class. The two lower maps (B) show the rural area around Wetaskiwin County in Alberta, and identify a considerable amount of change from Forest to Cropland. Both of these cases seem to be realistic. Although accumulation of error through comparison of individual input maps indicates that there could be as much as 30% error in the land use change statistics, it is apparent from the confusion matrices that error is biased toward specific classes and thus total error in change estimates may be no greater than the maximum for either of any two classes involved. However, it is difficult to estimate the actual error in such change results, and as a result we have initiated a ground-truth sampling study to assess and document the accuracy of land use change data derived from map comparison.

4. CONCLUSIONS

The study has shown that the development and application of rule sets to enable the integration of multiple raster input layers to produce high-resolution and high-accuracy land use maps is a realistic and successful approach, and it further shows that the accuracy of the output maps can surpass the accuracy of any of the inputs. It is also shown that accuracy increases as more input information is integrated.

We also feel that the incorporation of additional inputs such as slope, elevation, and soil maps could improve overall accuracy considerably, especially with respect to cropland, water, and wetland. By instituting simple rules such as "Wetland does not occur on sloping land" or "Cropland does not occur on shallow soils," accuracy could be further improved. It also seems apparent that the incorporation of unclassified Earth Observation data could improve accuracy as by, for example, eliminating the possibility of Grassland occurring on pixels with high normalized difference vegetation index (NDVI) values.

This study was somewhat primitive in the reliance on expert knowledge and human-developed rules for the resolution of conflicts between input data sources, but it seems that the integration of a variety of input sources could be much more expeditiously accomplished through the use of a single, simple rule of "highest accuracy." This would facilitate a much more automated and rapid approach, but would also rely on extensive, regionally specific accuracy assessments of input data. We are also investigating the potential of applying an artificial intelligence, machine-learning approach to the integration of multiple data sets.

In any case, the experience of this study indicates that to improve the accuracy of derived spatial data, new Earth Observation mapping efforts should build on existing maps rather than focus exclusively on new sensors and algorithms. By using existing information as a base, it appears that overall map accuracy could readily reach 95% or more.

ACKNOWLEDGMENTS

The authors would like to acknowledge the support of the Canadian Space Agency, Agriculture and Agri-Food Canada's National Carbon and Greenhouse Gas Accounting and Verification Study (NCGAVS), and National Agri-Environmental Health Analysis and Reporting Program (NAHARP) and the many individuals from Environment Canada, Natural Resources Canada, Statistics Canada, and Agriculture and Agri-Food Canada who supported and encouraged this undertaking.

REFERENCES

AAFC (Agriculture and Agri-Food Canada), 2009. Land Cover for Agricultural Regions of Canada, Circa 2000. Available from: http://open.canada.ca/data/en/dataset/16d2f828-96bb-468d-9b7d-1307c81e17b8.

AAFC, 2013a. National Ecological Framework. Available from: http://sis.agr.gc.ca/cansis/nsdb/ecostrat/index.html.

AAFC, 2013b. Annual Crop Inventory. Available from: http://open.canada.ca/data/en/dataset/ba2645d5-4458-414d-b196-6303ac06c1c9.

Friedl, M.A., Sulla-Menashe, D., Tan, B., Schneider, A., Ramankutty, N., Sibley, A., Huang, X., 2010. MODIS collection 5 global land cover: algorithm refinements and characterization of new datasets. Remote Sensing of Environment 114 (1), 168–182.

IPCC, 2006. Guidelines for National Greenhouse Gas Inventories, Volume 4: Agriculture, Forestry and Other Land Use; Chapter 3: Consistent Representation of Lands. Available from: http://www.ipcc-nggip.iges.or.jp/public/2006gl/vol4.html.

Kleeschulte, S., Bütt, G., September 2006. European Land Cover Mapping – The Corine Experience (Chapter 4) North America Land Cover Summit, Washington, DC. Available from: http://www.aag.org/cs/publications/special/nalcs.

Mayaux, P., Eva, H., Gallego, J., Strahler, A.H., Herold, M., Agrawal, S., Naumov, S., De Miranda, E.E., Di Bella, C.M., Ordoyne, C., Kopin, Y., Roy, P.S., 2006. Validation of the global land cover 2000 map. IEEE Transactions on Geoscience and Remote Sensing 44 (7), 1728–1737:1645273.

Natural Resources Canada, 2014. Canada Land Inventory (1:250,000) – 1966 Land Use. Available from: http://geogratis.gc.ca/api/en/nrcan-rncan/ess-sst/1cd7d940-309b-5965-b181-a5221dd898f3.html.

Natural Resources Canada, 2015a. Medium Resolution Land Cover Mapping of Canada from SPOT 4/5 Data. Available from: http://geoscan.nrcan.gc.ca/starweb/geoscan/servlet.starweb?path=geoscan/fulle.web&search1=R=295751.

Natural Resources Canada, 2015b. Geogratis. Available from: https://www.nrcan.gc.ca/earth-sciences/geography/topographic-information/free-data-geogratis/11042.

Natural Resources Canada, n.d. Canadian National Fire Database. Available from: http://cwfis.cfs.nrcan.gc.ca/ha/nfdb.

Ontario Ministry of Natural Resources, 2008. Southern Ontario Land Resource Information System (SOLRIS) Land Use Data. Toronto, Ontario Available from: https://www.javacoeapp.lrc.gov.on.ca/geonetwork/srv/en/main.home.

Pouliot, D., Latifovic, R., Zabcic, N., Guindon, L., Olthof, I., 2014. Development and assessment of a 250 m spatial resolution MODIS annual land cover time series (2000–2011) for the forest region of Canada derived from change-based updating. Remote Sensing of Environment 140, 731–743.

Statistics Canada, 2014. Census of Agriculture; Land Use. Available from: http://www5.statcan.gc.ca/cansim/a26?lang=eng&retrLang=eng&id=0040203&pattern=0040200..0040242&tabMode=dataTable&srchLan=-1&p1=1&p2=50#customizeTab.

Statistics Canada, 2016. Census of Agriculture. Available from: http://www.statcan.gc.ca/eng/survey/agriculture/3438.

University of Maryland, 1997–2016. Landsat GeoCover at the Global Land Cover Facility. Available from: http://glcfapp.glcf.umd.edu/research/portal/geocover/status.shtml.

Wickham, J.D., Stehman, S.V., Gass, L., Dewitz, J., Fry, J.A., Wade, T.G., 2013. Accuracy assessment of NLCD 2006 land cover and impervious surface. Remote Sensing of Environment 130, 294–304.

Wulder, M.A., Cranny, M.M., Hall, R.J., Luther, J.E., Beaudoin, A., White, J.C., Goodenough, D.G., Dechka, J.A., 2008. Satellite land cover mapping of Canada's forests: the EOSD land cover project (Chapter 3) In: Campbell, J.C., Jones, K.B., Smith, J.H., Koeppe, M.T. (Eds.), North America Land Cover Summit. American Association of Geographers, Washington, DC, USA, pp. 21–30. Available from: https://cfs.nrcan.gc.ca/publications?id=29220.

Chapter 6

France's Governmental Big Data Analytics: From Predictive to Prescriptive Using R

Henri Laude
Laboratoire BlueDxX, BlueSoft Group and Advanced Research Partners, Charenton le Pont, Paris, France

There cannot be a very large number of simple algorithms

A.N. Kolmogorov

1. INTRODUCTION

The goal of this chapter is to reach a better understanding of how agricultural practices and policies relevant to data science are articulated in France, to clarify the main features of an automated systems based on prescriptive analytics for agriculture.

Today, big data systems provide many descriptive statistics or predictive analytics that are not always easy to use. French farmers would like to improve their decision making and automate their processes without having to develop new computer skills. The best way to do this is mainly to provide farmers with prescriptive analytics or with automated systems based on prescriptive analytics.

Unsurprisingly, in terms of big data or smart data, the French open data and French open source landscape for data sciences are resolutely moving toward solutions based on proven technologies, such as languages R and Python, and Hadoop ecosystems (including SPARK, a general engine for large-scale data processing and streaming).

R is one of the most frequently used languages among data scientists in France, due in large part to its ability to invoke many machine learning packages (typically R, Python, Fortran, and C packages). Moreover, knowing that it is well integrated with the ecosystem evoked previously, we have focused our

Federal Data Science. http://dx.doi.org/10.1016/B978-0-12-812443-7.00006-5

study on the ability of R to easily cover the range of algorithms appropriate for the needs of data science applied to agriculture.

To do so, a taxonomy of the concepts linked to farming activities has been developed a well a mapping with the algorithms which are available in the R ecosystem.

Finally, this work outlines the main features of such automated systems based on prescriptive analytics and references existing components of the R ecosystem that could contribute to its implementation.

Owing to its deliberately minimalist nature, the approach has been dubbed "Parsimonious Modeling for Prescriptive Data Science."

2. MATERIALS AND METHODS: "PARSIMONIOUS MODELING FOR PRESCRIPTIVE DATA SCIENCE," APPLIED TO FRENCH AGRICULTURE

2.1 Agricultural Data in France

French open data initiatives are driven by a dedicated team named Etalab (Legifrance, 2015). Etalab conducts the policy of opening and sharing French public data under the authority of the French Prime Minister and is part of the General Secretariat for Modernization of Public Action.

The French Government's open data portal (www.data.gouv.fr, 2017) includes more than 19,000 data sets, which is 10 times less than the US Government's open data portal (www.data.gov, 2017). Nevertheless, it develops numerous initiatives to develop what it calls "The French Tech" around the use of these open data.

For example, the French graphic parcel register (RPG) is a geographic information system that allows the identification of the agricultural parcels. It consists of approximately 7,000,000 graphic objects, covering France and its overseas territories. The level of detail and the proactive process of updating such a data set show the willingness and capacity of France to engage in a new stage of its open data policy (Fig. 6.1) (Géportail, 2012).

RPG is one of the 33,000 data sets accessible on Geocatalogue, the French geographical repository built to be compliant with the "Infrastructure for Spatial Information in Europe" directive (UE, 2007). The INSPIRE directive establishes an Infrastructure for Spatial Information in the European Community and encompasses topics often related to agriculture: soil, agricultural and aquaculture facilities, atmospheric conditions, cadastral parcels, energy resources, geology, biotope, hydrography, land cover, economy, and base maps.

Many French companies advocate the application of big data in agriculture. These companies often focus on the link between precision agriculture (Cherkasov et al., 2009) and big data (Gartner Group, 2017). Invivo, one of the leading French cooperatives in the grain and agrosupply sectors, which announced a turnover of €6.4 billion, highlights: "big data applications will be

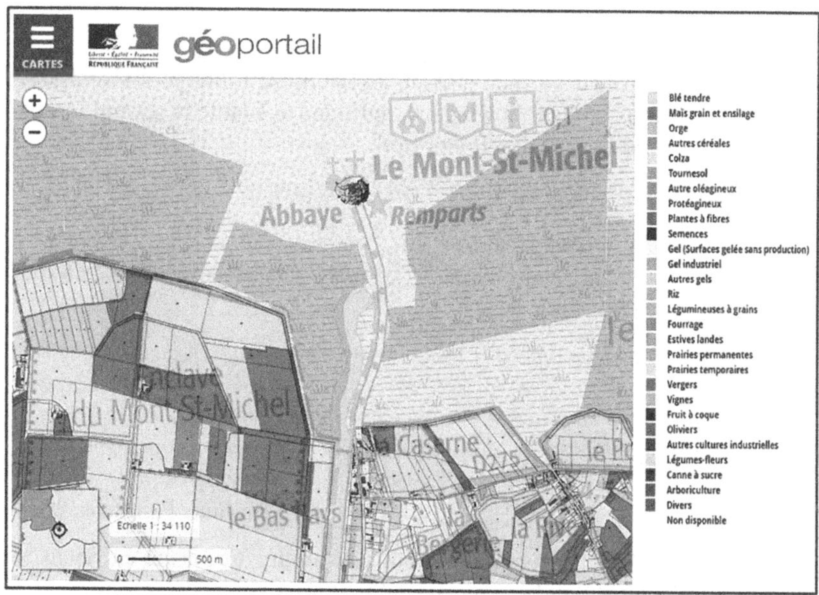

FIGURE 6.1 Detail of agriculture parcels around Mont-Saint-Michel, from RPG.

developed in sowing, crop protection, irrigation, the management of harvests, equipment, and livestock breeding" (Invivo, 2015).

These early findings provide some reassurance about the availability of data for agricultural data sciences. The consolidation of large data takes place in what we call a *data lake*.

Mastering the meaning of these data is a key factor, which permits an understanding of, and an appropriate response to, users' needs. Therefore the durability of such a data lake relies directly on taxonomies that will govern it (ontologies).

2.2 Open Taxonomies for Agricultural Data Sciences

Owing to the variety, the volume and the necessary reliability of data related to agriculture, building an agricultural data lake is a real challenge. European and French agricultural actors are aware of this reality. They contribute to projects such as AIMS (for Agricultural Information Management Standards), which "supports standards, technology and good practices for open access and open data in the agricultural domain." Under this umbrella, we find also the AGROVOC linked data repository, which provides links to other ontology repositories such as the STW Thesaurus for Economics and the Agricultural Thesaurus and Glossary (AIMS, 2017).

To construct a coherent and resilient taxonomy to structure a data lake, it is necessary to rely on existing taxonomies. Our reference thesaurus related to agriculture is published by the National Agricultural Library (USDA, 2017). There are records as in the following example, and available in several formats, including Resource Description Framework:

Entry: Alternative farming - EntryID: 7138 (simplified)

Links: ecological farming, fertilizer use reduction, pesticide use reduction, alternative crops, alternative livestock, organic production, reptile culture, sustainable agriculture, vermiculture.

Definition: Production methods other than energy and chemical-intensive one-crop (monoculture) farming. Alternatives include using animal and green manure rather than chemical fertilizers, integrated pest management instead of chemical pesticides, reduced tillage, crop rotation (especially with legumes to add nitrogen), alternative crops, or diversification of the farm enterprise.

Source: Agriculture Fact Book, USDA.

We can therefore be confident about our ability to master in detail the semantics of data that will be made available to data scientists. It remains to be seen whether these data scientists are in sufficient number, if their training is adequate, and if they use the appropriate tools (such as R).

2.3 Big Data and Data Science in France

The French authorities have established an ambitious road map for the development of big data: new training courses for data scientists, massive open online courses in French (MOOC), new ecosystems for start-ups, funding, open data, and new legislation (Hermelin and Bourdoncle, 2014). The CNLL (Conseil National du Logiciel Libre) points out that France has 30,000 employees in the open source sector (CNLL, 2013). According to a study conducted by IDC Research, Inc. in 2014, the French market of big data had a value of EUR 285 million. By 2018, the analyst firm expects an increase of 129% for a total value of EUR 652 million (IDC, 2016). In addition, Python and R are taught in the majority of scientific universities or schools at the post-baccalaureate level. Overall, data science ecosystem seems to be doing well and its future is looking bright. We can therefore have a look at the adequacy between the farming needs and the promises of French data science.

3. RESULTS

3.1 A Parsimonious Agricultural Taxonomy for Data Collection, an Intermediate Result

From the comparison and study of the database previously mentioned, this subsection proposes a minimal taxonomy tree (meaning with a minimal number of required classes). We found that seven classes were indeed sufficient to classify the abundant parameters relating to agriculture. Note that many topics could be combined together and that the corresponding attributes could then be irrelevant for a specific study. Many data should be spatially and temporally indexed.

Interested readers are invited to refer to the OLAP concepts (i.e., Online Analytical Processing hypercube used to analyze multidimensional data interactively) to better understand how it is possible to deal with the complexity and the multidimensional representation of the whole feature space (Aouiche et al., 2009).

Please note that to build our taxonomy, we have borrowed many concepts from the excellent report by the National Institute of Agricultural Research (INRA), the French public research institute dedicated to agricultural science (Guyomard et al., 2013), in which many key agricultural performance indicators are identified (i.e., they have been selected some of the 200 basic farming practices identified in INRA's work have been aggregated).

This subsection introduces a taxonomy that is in line with the goals of this chapter:

1. (FA) Farming activities
   a. Grain and oilseed crops/field crops
   b. Market gardening
   c. Arboriculture
   d. Viticulture
   e. Dairy cattle/milk
   f. Beef cattle
   g. Sheep and goat rearing
   h. Pig farming
   i. Poultry farming/egg production
2. (FP) Farming parameters
   Comment: The farming parameters to be collected, analyzed, and forecasted are many and varied. Numerous attributes should be collected for each topic (as appropriate: volumes, frequencies, lead times, storage costs, use by date, sell by date, purchase costs, operating costs, market price, over-the-counter price, resource availability, technical and quality key indicators, applicable current regulation, related key competencies, related actors, time stamp, geographic coordinates, other technical and scientific attributes).
3. (FP/PD) Farming parameters/Production data
   a. Plant protection products
   b. Pharmaceuticals
   c. Fertilizers
   d. Fuel and lubricants
   e. Water, electricity, and gas
   f. Selection and breeding
   g. Agricultural personnel

4. (FP/FI) Farming parameters/Financial and investment data
   a. Land prices and land rent
   b. Farm mechanization
   c. Farm building
   d. Agro-based food processing infrastructure (facilities or locally operated)
   e. Cash flow (actual and forecast)
   f. Operating subsidy
   g. Capital budget
   h. Allocation to depreciation and provisions/exchange rate
5. (FP/O) Farming parameters/Other technical data
   a. Soil organic matter content
   b. Geographical context. For example: is that farm located at 600 m above sea level (m.a.s.l.)?
   c. Clay content and soil texture
   d. Crop monitoring data
   e. Cation exchange capacity and soil pH
   f. Precipitation, temperature, evapotranspiration, wind speed and direction
   g. Soil fertility (dependency with other data)
   h. Regulatory and environmental norms and standards (now and tomorrow)
   i. Agricultural yields (commonly observed, past, actual, forecast)
6. (CRM) Customer relationship and market-related data
   a. Organoleptic properties
   b. Consumers habits
   c. Social networks data (including reputation, influencers, relationship)
   d. Institutional promotion to increase consumption (exist or not?)
   e. Market data
   f. Trading data
7. (PR) Day-to-day process management data

   Comment: At least main attributes should be collected for each topic (as appropriate: events related to the topic, transactions and counterparts, flows between entities, key indicators related to the management of the topic, status of each subprocess related to the topic).
   a. Financial management
   b. Contract and supplier management
   c. Pricing management
   d. Inventory management
   e. Labor management
   f. Greenhouse management
   g. Crop and livestock management
   h. Traceability, barcoding/radio-frequency identification management
   i. Agro-based food processing management
   j. Order processing
   k. Logistic and transport management
   l. Customer management

m. Innovation and R&D management
n. Data processing management
o. Crisis management (flood, plague, market shut down, etc.)

This minimal taxonomy is built in the spirit of the parsimony principle (as it is formulated within data science), i.e., "modeling with the smallest number of free parameters that explains the observed facts" (Vapnick, 2006). The subsequent work in this chapter relies on the taxonomy discussed earlier.

3.2 Agricultural Descriptive and Predictive Data Science With R

Using R language is a common practice in data science applied to agriculture. For example, see Meersmans et al. (2012) about an application on soil type and management.

Moreover, the Comprehensive R Archive Network (CRAN) gives many useful packages related to statistical inference, data manipulation, images processing, graphical visualization, time series prediction, spatial data, clustering, classification, and regression (Laude, 2016).

We also note that many multipurpose and powerful machine learning algorithms that are useful for farming data exploration are available within the R Caret package (Kuhn, 2016), including:

- K-means (clustering observations in k clusters)
- K-Nearest Neighbors (may be the best known classification algorithm)
- Boosted Generalized Linear Model (GLM) (an efficient generalization of the linear model to response variables that are not linear)
- Boosted Generalized Additive Model (efficient generalization of the GLM with response variables that could depend on functions of the predictors)
- CART (classification or regression tree, easy to interpret, numerical and categorical predictors)
- Regularized Random Forest (ensemble of decision trees with a safeguard against overfitting)
- Robust Quadratic Discriminant Analysis (the decision surface is a function of pairwise products of predictor, i.e., is not a hyperplan; "robust" means that outlier observations do not perturb the classifier)
- Naive Bayes Classifier (naive assumption regarding the independence of predictors/attributes, therefore high computing efficiency in term of scalability)
- Support Vector Machines (very efficient for binary classification, i.e., Y/N); because what it's called "the kernel trick" offers an easy transformation of the feature space, we can deal with very complex topologies without having to optimize many hyperparameters).

In relation with our taxonomy, specific use cases have been collected and the availability of R-related packages has been checked; examples are given in the following sections.

3.2.1 Crop Monitoring

Satellite or remote sensors crop monitoring technologies use large amounts of spatiotemporal data, images that evolve over time and whose interpretations are sometimes technically complex, especially because the frequencies used are often multispectral (for instance, visible and near-infrared). The corresponding algorithms available in CRAN include kriging, variogram estimation (Pebesma and Gräler, 2015), spatiotemporal parametric bootstrap, analysis of space–time series, linear mixed models, and nonlinear least squares problems as the Levenberg–Marquardt algorithm (Dong et al., 2016).

3.2.2 Evapotranspiration

The assessment of the evaporation and plant transpiration from the Earth to the atmosphere has been a major concern for agriculture for many decades. Please refer to Waller and Yitayew (2016) which is a good primer in crop evapotranspiration. The way to forecast the evaporation could be direct as in the R Evapotranspiration package (Danlu, 2016) or indirect by the use of machine learning algorithms as neural networks available in CRAN (Torres et al., 2011).

3.2.3 Prediction of Soil Fertility

For such a topic, we need a model that is not only effective, but also easy to interpret. That is the case, for instance, for the "Decision Trees and Rule-Based Models" because they are built on the values of the input attributes (Gholap, 2012). Note that the algorithm "C5.0" available in R is a more elaborate version than the "J48 (C4.5)" used for prediction of soil fertility by Jay Gholap in 2012.

The reader can conclude that the mapping between (1) the agricultural needs expressed through our taxonomy, (2) the descriptive or predictive analytics algorithms, and (3) the availability of R packages gives satisfactory results. Now it is time to move to the last step, and to check if R is a good candidate for prescriptive analytics.

3.3 From Descriptive Analytics to Prescriptive Analytics Through Predictive Analytics

3.3.1 Toward Prescriptive Analytics

Schematically, any decision taken on the basis of calculations on massive data or smart data (Lafrate, 2015) is presumed as involving three main steps: interpretation (descriptive), inference (predictive), and decision making (prescriptive).

Unfortunately, providing descriptive analytics to an end user will not drastically decrease his or her total workload related to these three steps (sometimes a farmer driving his tractor finds himself almost like a fighter pilot with too many sophisticated flight instruments; at other times the same farmer finds himself facing too many complex financial screens, and he should cope like a commodity trader on Euronext!).

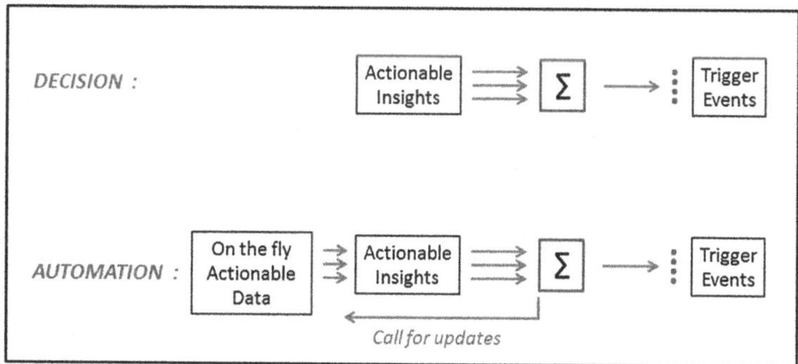

FIGURE 6.2 Aggregation engine (∑) in prescriptive analytics architectures.

Typically, descriptive analytics is what is most often found in companies under the misleading name of "Business Intelligence." Farmers are more efficiently helped when they are given clear predictions about the evolution of their environment and when they have the ability to simulate the impact of their decisions (i.e., not only with static graphics but also with tools to simulate and forecast; note that this type of tooling is the state of the art of the machine learning technologies implemented in big data architectures).

However, knowing that any prediction is uncertain and that several predictions must be combined to make a decision, the farmer's task can only be truly simplified if the corresponding actions are clearly prescribed. This is the purpose of prescriptive analytics or automated systems based on prescriptive analytics, often built on complex heuristics or artificial intelligences.

Roughly speaking, during this process the data goes through different states: raw data, actionable data (descriptive), actionable insights (predictive), and trigger events (prescriptive).

Consequently, we can deduce that the simplest architecture of a prescriptive analytics engine consists in **the aggregation (∑) of several actionable insights to generate a set of trigger events**. It is one of the key features of the upcoming generation of Decision Support System. In case you wish to automate a process, it is necessary to receive on the fly actionable data (Harth et al., 2013) continuously by requesting updates of the data upstream. Hence, it often makes sense to reinject the decisions taken and their effects (back propagation). Note that in many cases, automated processes in a big data environment are based on streaming technologies such as Apache-SPARK (Maarala et al., 2015) and that there is an Application Programming Interface ready for SPARK in Python and R (Fig. 6.2).

3.3.2 Three Main Kinds of Aggregation Engine

An aggregation engine could be any combination of these three main kinds of engine.

3.3.2.1 Ensemble Methods

Data scientists have a full set of models to aggregate several more basic learning models. These aggregation models are known as *Ensemble Methods*. Using Machine Learning Ensemble Methods is not always possible, because it often requires that the incoming models process the same information. For our specific problem, an Ensemble Method has to be chosen that does not require strong assumptions about the models to be aggregated. Therefore the practice that seems most attractive to us is to implement an energy analogy, through the implementation of a Boltzmann Machine (Villani, 2008), whose criterion of convergence is the stabilization of a measure of energy (which implies that it has the benefit of not prejudging the path taken to converge).

3.3.2.2 Optimization of a Utility Function

Searching the hyperparameters of the aggregation engine can be done by implementing an optimization process on a utility function. Such a utility function is not easy to construct because it must have an unambiguous meaning for users of the system (e.g., like a financial utility function). Owing to the finite number of data samples collected in farming activity, and at the opposite because of the high-dimensional feature space, we have to deal with sparse arrays. That is why the classical optimization algorithms such as Gradient Descent are not efficient.

Therefore we have adopted the Particle Swarm Optimization algorithm (PSO), based on an analogy with the behavior of a swarm of insects looking for food (Clerc, 2006). Note that PSO is a better choice given that it can be adapted to multiobjectives problems (i.e., multiple utility functions in parallel).

3.3.2.3 Heuristic and Fuzzy

Heuristic approaches have a broader scope but are difficult to validate, typically when we capture and model the knowledge of various experts in a set of rules. Most of the time these rules are expressed with explicit uncertainties (e.g., "if there is such a symptom, I think it is quite possible that"). There are two common ways to deal with this type of fuzzy knowledge, when appropriate:

1. When experts express the rules in a quantitative form, then data scientists often transform the problem of aggregation as an FLP (Fuzzy Linear Programming Problem) that consists of maximizing or minimizing a function subject to linear constraints, but having imprecision in both the coefficients and the constraints (Li, 2014).
2. When experts express the rules as *if-then-else* clauses, then we have to use a Fuzzy Rules Based System (FRBS). An FRBS looks like a classical Expert System (Vizureanu, 2010) but has the ability to manage fuzzy numbers or sets (Li, 2014).

3.3.3 Available R Building Blocks for Building an Aggregation Engine (Nonexhaustive Mapping)

Several building blocks could be parts of an Aggregation Engine, so several packages from the CRAN repository have been selected and tested. The result of this work is the following proposed *starter kit*, which allows one to effectively deal with the functional requirements related to the three main kinds of aggregation engine. These packages can be used together to build an efficient aggregation engine. The six packages of this starter kit are:

1. Packages to build "Ensemble Methods"
   Package_1: *caretEnsemble*, "Functions for creating ensembles of caret models" (Deane-Mayer et al., 2016).
   Package_2: *deepnet::rbm.train*, "Training an RBM (Restricted Boltzmann Machine)" (Rong, 2014).
2. Packages to build and optimize a utility function
   Package_3: *utility*, "Construct and plot objective hierarchies and associated value and utility functions" (Reichert et al., 2013).
   Package_4: *psoptim*, Particle swarm optimization (Krzysztof, 2016).
3. Packages to manage heuristic and fuzzy behaviors
   Package_5: *FuzzyLP*, "Fuzzy Linear Programming Problems with fuzzy constraints, fuzzy costs and fuzzy technological matrix" (Villacorta et al., 2015).
   Package_6: *fbrs*, " learning algorithms based on fuzzy rule-based systems (FRBSs) for dealing with classification and regression tasks. Moreover, it allows construction of an FRBS model defined by human experts" (Riza et al., 2015).

The preceding facts and observations, and in particular those of Sections 3.2 and 3.3.2, support the idea that **prescriptive analytics based on an R ecosystem** can fulfill the needs of agricultural prescriptive analytics research goals.

4. CONCLUSION

The agricultural practices and the policies related to data sciences in France are healthy:

- availability of data to build efficient data lakes that can rely on up-to-date taxonomies
- data scientists in sufficient number, with adequate training
- dynamic ecosystem for big data and R
- availability of R packages for many needs related to our taxonomy
- state-of-the-art descriptive and predictive analytics practices.

Taking all this into account, this chapter has proposed a new framework, the "Parsimonious Modeling for Prescriptive Data Science," applied to the French agriculture needs. Using this "Parsimonious Modeling for Prescriptive Data Science" method, a taxonomy of farming and the schema of a new architecture for prescriptive analytics for the French agriculture have been proposed.

This new hybrid taxonomy combines financial and farming topics to support business and production analytics. To ensure that French farmers can improve their decision making and automate their processes by the use of prescriptive analytics systems, we have defined the key components of a prescriptive analytics architecture, around the concept of aggregation engine.

We have put forward a starter kit for building such an aggregation engine with available R packages from the CRAN repository. The starter kit encompasses the following aspects:

- Ensemble methods and Boltzmann machines training
- Optimization of utility functions through Particle Swarm Optimization
- Managing heuristic and fuzzy behaviors through fuzzy linear programming or learning algorithms based on fuzzy rule-based systems.

Future work should enrich the starter kit with other packages. The main efforts will be devoted to work on specialized algorithms for farming and commodity trading.

REFERENCES

AIMS, 2017. Linked Data | Agricultural Information Management Standards (AIMS). Available at: http://aims.fao.org/aos/agrovoc/.

Aouiche, K., Lemire, D., Godin, R., 2009. Web 2.0 OLAP: from data cubes to tag clouds. Lecture Notes in Business Information Processing, 18 LNBIP 51–64.

Cherkasov, G.N., Nechaev, L. a., Koroteev, V.I., 2009. Precision agricultural system in modern terms and definitions. Russian Agricultural Sciences 35 (5), 334–338. Available at: http://www.springerlink.com/index/10.3103/S1068367409050139.

Clerc, M., 2006. Particle Swarm Optimization. ISTE Ltd.

CNLL, 2013. panorama-open-source-2013-CNLL. Available at: http://www.entrepreneursdavenir.com/flw/article/-/id/1005.

Danlu, G., 2016. Package "Evapotranspiration" Title Modelling Actual, Potential and Reference Crop Evapotranspiration. Available at: https://mran.microsoft.com/web/packages/Evapotranspiration/Evapotranspiration.pdf.

Deane-Mayer, A.Z.A., Knowles, J.E., Deane-Mayer, M.Z.A., 2016. Package "caretEnsemble," the Comprehensive R Archive Network – CRAN.

Dong, J., et al., 2016. 4D crop monitoring: spatio-temporal reconstruction for agriculture. In: IEEE International Conference on Robotics and Automation Available at: http://arxiv.org/abs/1610.02482.

Gartner Group, 2017. What Is Big Data? – Gartner IT Glossary – Big Data. Available at: http://www.gartner.com/it-glossary/big-data.

Géportail, 2012. Registre Parcellaire Graphique (RPG) 2012-Géoportail. Available at: https://www.geoportail.gouv.fr/donnees/registre-parcellaire-graphique-rpg-2012.

Gholap, J., 2012. Performance tuning of J48 algorithm for prediction of soil fertility. Asian Journal of Computer Science and Information Technology 2 (8), 251–252.

Guyomard, H., et al., 2013. Vers des agricultures à hautes performances. INRA 4, 1–40. Available at: http://institut.inra.fr/Missions/Eclairer-les-decisions/Etudes/Toutes-les-actualites/Rapport-Agricultures-hautes-performances#.

Harth, A., et al., 2013. On-the-fly integration of static and dynamic linked data. In: CEUR Workshop Proceedings, vol. 1034, p. 257641.

Hermelin, P., Bourdoncle, F., 2014. Big Data – Feuille de route comité de pilotage des plans de la Nouvelle France Industrielle, p. 40.

IDC, 2016. [Infographie] IDC : enjeux et dynamique du Big Data en France. Available at: http://www.lebigdata.fr/infographie-idc-enjeux-dynamique-big-data-france.

Invivo, 2015. Big Data et Agriculture – YouTube. Available at: https://www.youtube.com/watch?v=j28djIUN_6o.

Krzysztof, C., 2016. Package Psoptim, the Comprehensive R Archive Network – CRAN.

Kuhn, M., 2016. The Caret Package. Available at: https://topepo.github.io/caret/index.html.

Lafrate, F., 2015. From Big Data to Smart Data. ISTE Ltd and John Wiley & Sons, Inc.

Laude, H., 2016. Data scientist et langage R : guide d'autoformation à l'exploitation des big data ENI. Available at: http://www.editions-eni.fr/supports-de-cours/livre/data-scientist-et-langage-r-guide-d-autoformation-a-l-exploitation-des-big-data-9782409000430.

Legifrance, 2015. Arrêté du 21 septembre 2015 portant organisation du secrétariat général pour la modernisation de l'action publique | Legifrance. Available at: https://www.legifrance.gouv.fr/eli/arrete/2015/9/21/PRMX1516546A/jo/texte.

Li, D.-F., 2014. Decision and Game Theory in Management with Intuitionistic Fuzzy Sets. Springer International Publishing.

Maarala, A.I., et al., 2015. Low latency analytics for streaming traffic data with Apache Spark. In: Proceedings – 2015 IEEE International Conference on Big Data, IEEE Big Data 2015, pp. 2855–2858.

Meersmans, J., et al., 2012. A novel soil organic C model using climate, soil type and management data at the national scale in France. Agronomy for Sustainable Development 32 (4), 873–888.

Pebesma, E., Gräler, B., 2015. Introduction to Spatio-Temporal Variography, pp. 1–11 Available at: https://cran.r-project.org/web/packages/gstat/vignettes/st.pdf.

Reichert, P., Schuwirth, N., Langhans, S., 2013. Constructing, evaluating and visualizing value and utility functions for decision support. Environmental Modelling and Software 46, 283–291.

Riza, L., et al., 2015. FRBS, R Fuzzy Rule Based System, the Comprehensive R Archive Network – CRAN. Available at: http://sci2s.ugr.es/dicits/software/FRBS.

Rong, X., 2014. Deepnet: Deep Learning Toolkit in R, the Comprehensive R Archive Network – CRAN.

Torres, A.F., Walker, W.R., McKee, M., 2011. Forecasting daily potential evapotranspiration using machine learning and limited climatic data. Agricultural Water Management 98 (4), 553–562. Available at: http://dx.doi.org/10.1016/j.agwat.2010.10.012.

UE, 2007. Directive 2007/2/EC. Available at: http://eur-lex.europa.eu/legal-content/EN/ALL/?uri=CELEX:32007L0002.

USDA, 2017. National Agricultural Library, Thesaurus. Available at: https://agclass.nal.usda.gov/about.shtml.

Vapnick, V., 2006. In: Jordan, M., Kleingerg, J., Schölkopf, B. (Eds.), Information Science and Statistics. Springer International Publishing.

Villacorta, P.J., et al., 2015. FuzzyLP: An R Package for Solving Fuzzy Linear Programming Problems, the Comprehensive R Archive Network – CRAN.

Villani, C., 2008. Entropy production and convergence to equilibrium. In: Entropy Methods for the Boltzmann Equation. Springer International Publishing, pp. 1–70.

Vizureanu, P., 2010. In: Vizureanu, P., Smiljanic, T. (Eds.), Expert Systems. Intech, Vukovar, Croatia.

Waller, P., Yitayew, M., 2016. Crop evapotranspiration. In: Irrigation and Drainage Engineering. Springer International Publishing, pp. 89–104.

Chapter 7

Agricultural Remote Sensing and Data Science in China

Zhongxin Chen[1], Haizhu Pan[1], Changan Liu[1], Zhiwei Jiang[2]
[1]*Institute of Agricultural Resources and Regional Planning, Chinese Academy of Agricultural Sciences, Beijing, China;* [2]*National Meteorological Information Center of China, Beijing, China*

We show governments across the world that it makes financial sense to release broadly and openly the data that the taxpayer has already paid for. It will build your economy, it is certainly good for transparent government and, it is a boon for capacity building and education.

Barbara Ryan, Secretariat Director of the intergovernmental Group on Earth Observations (GEO)

1. AGRICULTURAL REMOTE SENSING IN CHINA

1.1 Agricultural Remote Sensing Research and Applications

In China, the research and applications of remote sensing technology in agricultural monitoring began in 1980s. Owing to the small size of farmland, diversity of crop species/variety, complicated cropping patterns, and the geomorphological complexity of cropland in China, the conventional in situ investigation method is greatly influenced by human factors, which is time consuming, expensive, and difficult to adapt to the needs of decision making of government departments. Since 2000, China's agricultural production has suffered intensive change. Get the dynamic, large-scale, and timely spatial information of crop production process is crucially needed, through remote sensing technology. Remote sensing technology has the advantages of quick and nondestructive acquisition of agricultural information; its rapid development can provide the necessary information for agricultural production process management. In China, the research and applications of remote sensing in agriculture mainly involves the radiation transmission mechanism of cropland and crop parameter remote sensing inversion, crop remote sensing classification, crop yield and quality prediction, agricultural disaster and forecasting, and agricultural remote sensing data acquisition technology.

Federal Data Science. http://dx.doi.org/10.1016/B978-0-12-812443-7.00007-7

Agricultural remote sensing information acquisition is the basis of agricultural remote sensing application. In the past, China's agricultural remote sensing data relied on foreign satellites. Nowadays, the domestic satellite remote sensing, aerial remote sensing Unmanned Aerial Vehicles (UAV), and ground close range remote sensing are widely applied. A collaborative technical system of agricultural remote sensing information acquisition is established. The successful launch of the China–Brazilian Earth Resource Satellite (CBERS-1) in 1999 marked the new era that the agricultural remote sensing community have independent land resource satellite remote sensing data sources in China. CBERS data were widely used in crop area estimation, growth monitoring, and pest and disease remote sensing (Jiang et al., 2001). The small satellites of HJ-1 A/B, Beijing-1, and Beijing-2 provided more domestic remote sensing for agricultural remote sensing data. Especially the series of China high-resolution earth observation system (CHEOS, or Gaofen, GF) satellites will provide all-weather, high-resolution, hyperspectral, short-revisit, and global-coverage remote sensing data for agriculture and other sectors. Launched in 2013, the first CHEOS satellite, namely GF-1, has provided high-quality remote sensing data for agriculture. The GF-1 WFV (wide-field viewer) sensor, with the spatial resolution 16m and revisit 4-5d, provides the ideal data for crop mapping, crop growth, agricultural environment and disaster monitoring, and so on. There is a 4-m multispectral scanner and 2-m panchromatic camera on board GF-1, which can be used in precision farming and agricultural engineering. There are seven satellites in the CHEOS; four of them (GF-1, GF-2m, GF-3, and GF-4) have been launched. In China, UAVs have been extensively used in agriculture monitoring in the past 5 years, because UAV is inexpensive to retrieve remote sensing data at farm and regional scales. It has several advantages in multiphase data acquisition. For agriculture, UAVs have been used for fine geometric measurement, fine monitoring of the crop height, distribution, density, growth, stress, and health. The UAV hyperspectral image analysis technique has been used in crop diseases and soil fertility monitoring and diagnosis. The information can be used to enhance the efficiency of fertilizer and pesticide application. An integrated farmland information acquisition system based on satellite, UAV, and ground Internet of things was established in China. Shi and colleagues proposed a technology framework of satellite, UAV, and ground remote sensing data registration, and the key technologies of crop information acquisition, and improved the capability of farmland information acquisition (Shi et al., 2014).

In-depth understanding of the soil-leaf-canopy radiation transfer process plays an important role in the quantitative remote sensing crop monitoring and evaluation. Many important advances have been made in the radiation transfer models of crops in China. Yang developed a farmland-atmospheric-sensor radiation transfer simulation system by coupling the leaf-soil-canopy radiation transfer model with the numerical analysis of atmospheric radiation transmission model and the sensor imaging model (Yang et al., 2013). The system has realized the multitemporal, multiangle, hyperspectral, and high-resolution remote sensing data simulation. It laid a foundation for agricultural quantitative remote sensing research and applications. Agricultural quantitative remote sensing is an important way to obtain the key biochemical parameters. It can provide valuable reference information for the field agricultural management, and can also be used as the input data for the crop growth model, crop yield estimation, and data assimilation. The main remote sensing inversion of crop biochemical parameters include Leaf Area Index (LAI), canopy chlorophyll content, above-ground biomass, and leaf water content. At present, there are two categories of inversion methods. One is the empirical statistical model based on the relationship between crop parameters and spectral reflectance or vegetation indices. The other is the inversion method based on the radiation transfer model. LAI is one of the hot topics of crop quantitative remote sensing. LAI has significance for crop transpiration, photosynthesis, and crop yield estimation. It is also an important parameter of the biogeochemical cycle and the hydrothermal cycle between land and atmosphere. Since the mid-1990s, quantitative remote sensing inversion of LAI has been developed rapidly in China. The data sources vary from the single multispectral data to the hyperspectral data, multiangle data, radar data, high-resolution data, and UAV data. Especially, the domestic satellite data have got increasing attention in quantitative inversion of LAI, such as HJ-1 A/B and GF-1 data (Jia et al., 2015). The LAI inversion research involved the common crops (wheat and maize) and other crop products (sugar cane, cotton, tobacco, etc.) (He et al., 2013; Wang et al., 2015). Soil moisture is the important factor in the crop growth process. Soil moisture monitoring data are very important for agricultural irrigation, crop photosynthesis and nutrient uptake, and yield prediction. Chinese researchers have used remote sensing to the regional soil moisture inversion. The thermal inertia index, temperature vegetation index, and vegetation supply water index are the main methods applied in regional soil moisture inversion (Song et al., 2011; Yang and Yang, 2016). In addition to the LAI and soil moisture, quantitative remote sensing inversion research in other important field environment parameters have also been committed. For vegetation biochemical composition inversion, the research were carried out on the mechanism of detecting leaf chemical components by remote sensing image spectrometry (Niu et al., 2000). For other important vegetation biochemical components, such as leaf water content, photosynthetically active radiation, and biomass, several studies have been carried out for the agricultural quantitative remote sensing both in the empirical statistical approach and in the physical model approach in China (Tian et al., 2000; Xue et al., 2005; Feng et al., 2008). Cropland classification and crop mapping is another important aspect of agricultural remote sensing research. It is the basis of monitoring crop acreage, growth, yield, and disaster. The main crop remote sensing classification and identification methods are based on the crop information of spectral characteristics, texture features, phenological features, and agronomic mechanism. The combination of optical and microwave data for crop classification has also received considerable attention. Although multispectral remote sensing is the main data source, much higher spatial resolution data and hyperspectral data are used in China.

The remote sensing applications in agriculture include agricultural survey, crop yield, agricultural disaster forecast, and precision agriculture. For different applications, different spatial resolution optical and microwave remote sensing data have different advantages. For example high-spatial resolution data are mainly applied in the precision agricultural at field scale, whereas high-temporal resolution data are mainly used for crop growth monitoring. A quick and accurate estimation of China's major crop acreage and its spatial distribution can assist the government departments to formulate a scientific food policy. In the 1980s, the China Meteorological Administration (CMA) carried out winter wheat monitoring using satellite data in the northern 11 provinces in China. In the 1990s, the Chinese Academy of Science used the Landsat/TM to monitor the wheat, rice, and maize planting area in China. After 2000, time series EOS-MODIS remote sensing data were used in crop acreage estimation (Xiao et al., 2005). With the successful launch of the GF series satellites, more high-resolution data were used in the cropland monitoring. The Remote Sensing Application Center of the Ministry of Agriculture of China carried out crop mapping for wheat, maize, rice, and soybean with GF-1 and other satellite data. GF satellite data have been the main data source in the Chinese agricultural remote sensing operation. The status and trend of crop growth directly affects the final yield and quality of the crop. Crop growth remote sensing monitoring provides the basic information for agricultural production management, and crop yield prediction. The crop growth monitoring methods include direct statistical monitoring, the inter-annual comparison, and crop growth process monitoring methods (Chen et al., 2000; Wu et al., 2004; Yang and Pei, 1999). The statistical monitoring methods are mainly based on remote sensing vegetation index to obtain crop growth monitoring results (Huang et al., 2012). The inter-annual comparison method mainly uses the difference or ratio of inter-annual remote sensing index to crop growth monitoring. This method is the main method of crop growth monitoring; it was widely used in the CropWatch system established by the Chinese Academy of Science and rice remote sensing yield estimation system established by Zhejiang University (Chen and Sun, 1997; Wu et al., 2010, 2014). The crop yield remote sensing estimation models include empirical models, semimechanism models, and mechanism models. In recent years, the crop yield simulation study by comparison of different crop models (WOFST, DSSAT, EPIC, etc.) and different data assimilation methods (EnKF, PF, SCE-UA, etc.) was carried out in China (Ma et al., 2004; Fang et al., 2008; Jiang et al., 2012).

China is a country with a high incidence of agricultural natural disasters; the average annually affected area accounts for 31.1% of the crop planting area. Since the 1980s, China started the study on remote sensing application of agricultural drought, flood, and disease disaster (Ju and Sun, 1997). Since the 2000s, regional and national agricultural disaster remote sensing monitoring systems have been set up and widely used (Yan et al., 2006; Yang, 2013).

In the past 20 years, agricultural remote sensing research and applications have made great progress in China. During 2016–2020, with the further implementation of the high-resolution earth observation project and the national space infrastructure, there will be more domestic satellites launched. At the same time, with the development of sensors, Internet of things, big data, and artificial intelligence, agricultural remote sensing research and applications will be further developed.

1.2 China Agricultural Remote Sensing Monitoring System

In the 1990s, several operational crop remote sensing monitoring systems have been set up and put into operation based on the achievements and outcome of several previous research projects since the 1980s. With the rapid development of earth observing instruments and technology, there has been increasing concerns regarding the monitoring of agriculture with remote sensing in China since the late 1990s. Several departments and research agencies have intensive research on it, whereas some of them set up their independent remote-sensing-based crop or agriculture monitoring systems. Among these operational systems, Ministry of Agriculture (MOA) China Agriculture Remote Sensing Monitoring System (CHARMS), Chinese Academy of Sciences (CAS) China CropWatch system, and CMA crop growth monitoring and yield prediction system are the major ones.

CHARMS was developed by the MOA Remote Sensing Application Center based on previous extensive related domestic and international research (Chen et al., 2011). It has been operational since 1999. It monitors crop acreage change, yield, production, crop growth, drought, and other agro-information for seven main crops in China, including wheat, rice, maize, soybean, cotton, canola, and sugarcane. The system of CHARMS consists of the following components: the database subsystem, crop acreage change monitoring module, crop yield estimation module, crop growth monitoring module, soil moisture monitoring module, disaster monitoring module, and information service module (refer to Fig. 7.1). It supplies the monitoring information to MOA and related agriculture management sectors in the form of ad hoc reports according to MOA's Agriculture Information Dissemination Calendar, with more than 100 reports per year. In China, the agricultural remote sensing monitoring system has become a mature and highly operational system.

2. DATA SCIENCE IN CHINA

2.1 Data Science Development in China

With the rapid development of information technology, worldwide business is experiencing information changes, and almost every industry is trying to exploit and utilize the value of big data. According to the International Data Corporation (IDC), all the digital data created, replicated, and consumed will double every

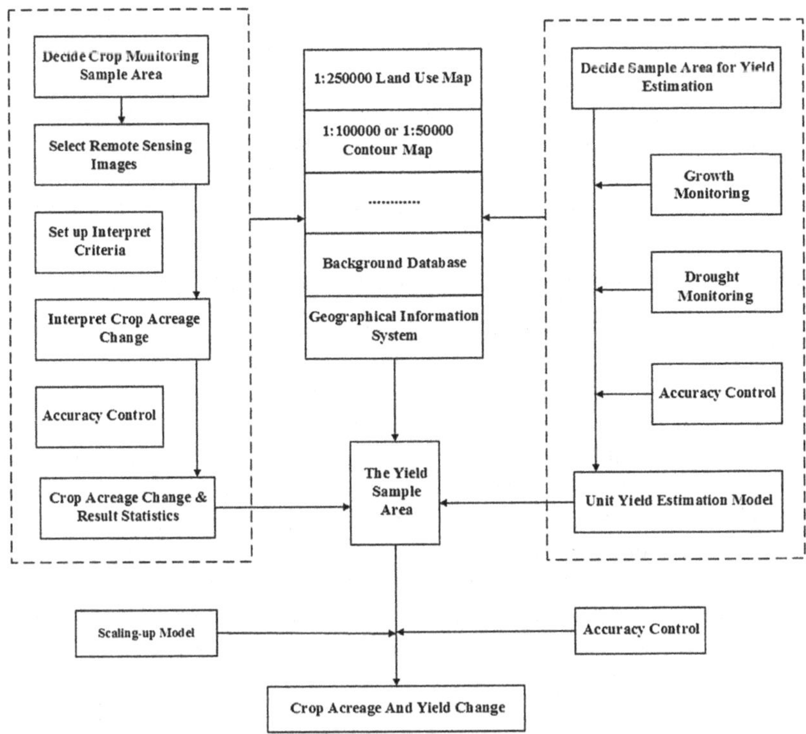

FIGURE 7.1 The structure of China Agriculture Remote Sensing Monitoring System.

2 years until 2020 (Gantz and Reinsel, 2012). The IDC estimates that by 2020, the data will reach 40 ZB, with China's global share increasing from 13% in 2012 to 21%. Data science as one of the most basic and influential resources of scientific and technological innovation in the information age has a significant ability to promote science and technology, application of value-added potential, and decision support. Therefore it is an important part of technological innovation basis platform. Data science became an important strategy to promote industrialization, now it has existed in every social field in China.

Since the 1970s, China has carried out various observation, investigation, and experimental research in various scientific fields. A huge volume of scientific data has been accumulated. These scientific data have provided important basic data and operational basis for the creation of innovative scientific theories, the exploitation and utilization of resources, and the construction of major projects and environmental protection, and also became the basis of economic development and social progress. According to incomplete statistics, only in the 1996–2000 period the investment was more than 50 billion Yuan in the field of resources and environment data collection. Over the same period, nearly 200 billion Yuan was invested in the state-level science and technology programs, 50% of which was used for the collection and collation of basic scientific data (Xu, 2003). So far, China has successively built a number of national scientific data centers, covering the basic content of science and technology in various fields. The relevant ministries have also set up a specific information center, responsible for the collection and sorting of the various types of data, through the establishment of the database and product processing, to provide users with information services and the corresponding technical support. At present, these information centers have become the government departments' important portals to provide public welfare and basic services. At the same time, China's science researcher also used the digital means to collecting data. A large number of scientific databases has been established; many database projects have been carried out for data service. The development of database in China has gone through three periods (Wu, 1995). The first period is the learning stage from the 1970s to 1980s, when we learned foreign successful experiences from US and other developed countries. The second period is the development stage from the 1980s to 1990s, when we started to construct the database. The third period is the stable development stage, from the 1990s to the present, when we began improving the quality of the database. In 1992, the number of big databases in China was only 137; however, the number reached 1038 in 1995. After 2000, with the development of information technology and network, China's database construction and application has begun to accelerate (Wang, 2009). E-government was listed as an important task in the 2000–2005 5-year period. The large amounts of funds put into the development of e-government attracted the participation of many developers and information technology companies. The government website has been greatly improved in content and function, and the government online has entered the practical stage.

The Ministry of Science and Technology (MOST) started a number of technology data science research projects to support the construction of a number of scientific databases through scientific and technological programs (Sun et al., 2013; Wen, 2013). The China Bureau of Meteorology, with the support from MOST, started a meteorological database development and data sharing in 2001. A lot of "golden" projects were committed in different ministries to improve database setup and information sharing. In the Ministry of Agriculture, the Golden Farm Project was put forward with the aim of accelerating and advancing agricultural and rural information and establishing the agricultural integrated management and services system (Du, 2008). The main tasks include the network and information services, establishment and maintenance of national agricultural database group and its application system, coordinate the development of unified information collection, the organization of agricultural modernization information service, and the promotion of various types of computer applications, such as geographic information system and remote sensing information system. Data science has developed in the government and the scientific research field; in addition, it has rapidly

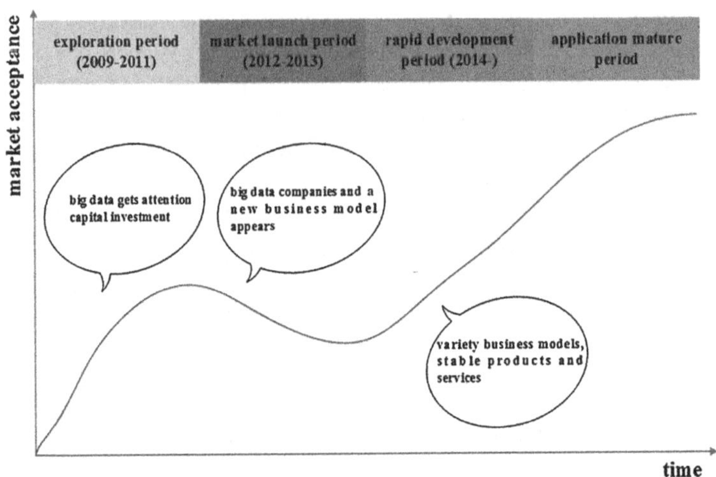

FIGURE 7.2 Big data industry evolution in China.

developed in the enterprise, financial, and other fields, and formed the big data industry (refer to Fig. 7.2). Now big data in China is in a period of rapid development. Big data has become the important strategy for many companies, such as Baidu, Alibaba, Tencent, JD.com, and other internet companies. Big data is evolving from theoretical research to practical applications, from the professional fields to industrial domains. Ali cloud as a well-known cloud computing platform, serves in the financial, electricity providers, logistics, television, and other industries.

With the rapid increase in the amount of remote sensing data, remote sensing data science gets more and more attention. Since the 1980s, China has launched a series of meteorological, oceanic, resources, and environmental satellites. Remote sensing has been extensively applied in various sectors in China. During the 2010–2015 period, China started the "Satellite-Airplane-Ground" integrated quantitative remote sensing system and applied demonstration project, under the National High Technology Research and Development Program (863 Program). It constructed the national remote sensing network to enhance the spatial information service capability of data receiving, processing, quantitative remote sensing, and application. With the implementation of the national major project "High-resolution Earth Observation," it began to develop GF satellite and put it into use from 2013, and build the system around 2020. To observe and understand the water cycle system, the Water Cycle Observation Mission program is proposed and is expected to be launched around 2020. With the development of observation technology, in recent years, the remote sensing data are also in explosive growth in China and the remote sensing field has stepped into the remote sensing big data age (Zhu et al., 2016).

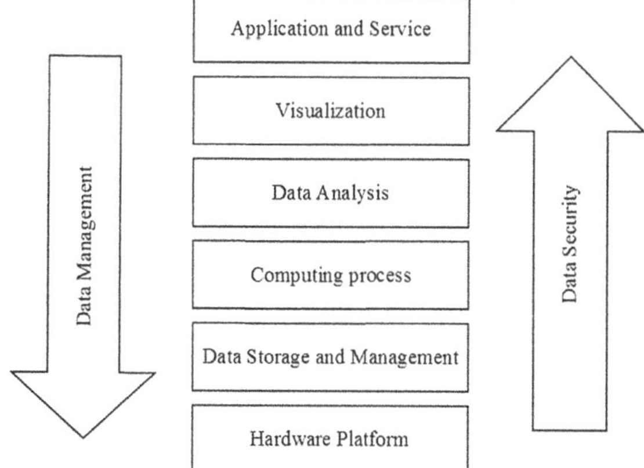

FIGURE 7.3 Big data technology process.

Data science not only is massive data, but also has the ability to process and analyze, and dig the value of the data (Liu and Zhang, 2014). Fig. 7.3 shows a typical big data technology process. It cannot be separated from the computer technology, hardware, and network development. The key technology of big data includes distributed clustered storage, high-performance computing, data mining algorithm, and big data visualization. During the 2010–2015 period, the data cloud of the CAS formed a multilevel and cross-type information service system with the infrastructure cloud service, the research data cloud service, and the data application cloud service as the main body. With the help of the cloud computing system, the CAS Computer Network Information Center provides support for the research projects. As the development of bioinformatics molecular data analysis environment, geospatial data cloud, data visualization and other application, multiple areas have formed the big data applications and service models (Han and Cercone, 1999).

2.2 Science Data Sharing and Services Platforms

If the data-driven information collection is a significant feature of data science research, then data sharing is the basis for big data application. The data sharing platform is based on the management of the data set and provides data and data services to users. In 1982, the CAS put forward the scientific database and information system projects, which has become a comprehensive scientific information service system after 20 years' development. In 1989, the CAS united relevant departments and scientific research institutions to set up the China Data Center of World Data Center (Zhu et al., 2010). In 2001,

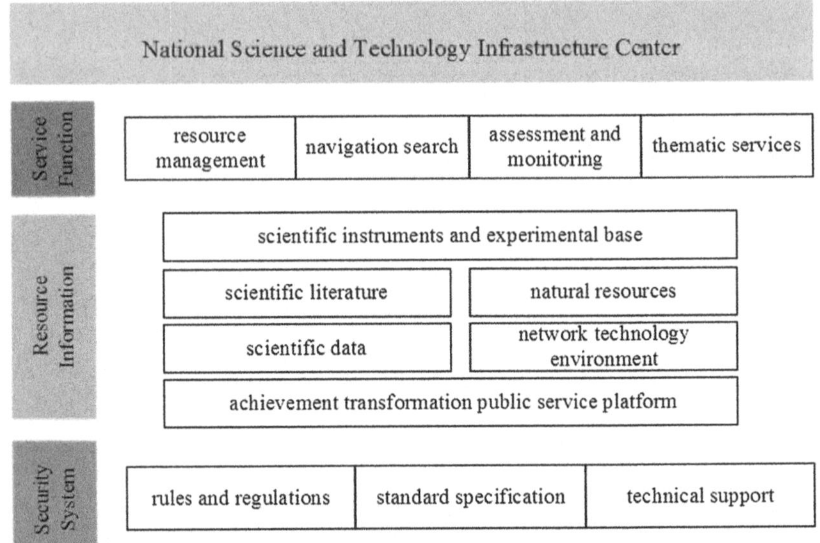

FIGURE 7.4 The structure of NSTIC.

MOST published the research report "Implement scientific data sharing project to promote scientific and technological innovation," which investigated the main problems and possible solutions in the current scientific data sharing. The Ministry of Science and Technology, Chinese Academy of Sciences, and other departments established a scientific data sharing and service platform through the project investment.

In China, data sharing in recent years has made significant progress; however, there is still some lag from that of the developed countries in terms of the degree of sharing, influence, the depth of research on data science, and the advancement of technology. The main problems in China science data sharing are decentralized investment and management, waste of scientific and technological resources, sector closure, information retention, and data monopoly. To resolve these problems, and to strengthen the scientific and technological innovation capacity building, the Chinese government started to organize the science data sharing and service network during 2005–2010 (Liu, 2005; Wu, 2004), and founded the National Science and Technology Infrastructure Center (NSTIC). The NSTIC is the basic supporting system for serving the innovation for the whole nation. Fig. 7.4 shows the structure of the NSTIC. The NSTIC is a jointly built and shared, rational layout, full-featured, open and efficient information security system, using the information technology to reorganize and optimize large-scale scientific and technological facilities, natural science and technology resources, scientific literature, science and technology basic data, and other basic conditions resources. NSTIC is used in government science and technology resource allocation activities. It consists of large scientific instruments and experimental base construction, natural science and technology resources sharing, scientific data sharing, literature sharing and so on. The scientific data sharing service center includes the forestry scientific data platform, earth system science data sharing infrastructure, population and health science data sharing platform, agricultural science data sharing center, earthquake science data sharing center, and meteorological science data sharing center. At present, the scientific data sharing platform has been basically completed, and has started providing data service in various fields. For example, in 2001 the CMA as the first batch of experimental units in China's scientific data sharing project provided free meteorological information for scientific research institutions. Since 2015, the China meteorological data network has provided official online service. By the end of 2016, CMA meteorological data center online access amount has been up to 130 million people times with over 100,000 registered users. The total amount of shared data for users in various fields is more than 800 TB. At the same time, the National Earth System Science Data Sharing Infrastructure (National Science and Technology Infrastructure, 2014) as the important earth science data sharing platform has provided the effective data services for the scientific research and social projects. By the end of 2014, the platform had 59.56 TB of data resources, and provided 93.53 TB of data services to the scientific research and the general public. During the same period, many universities began to build specific science data sharing and service platforms, such as the research center of remote sensing big data in Tsinghua University and the agricultural big data cloud platform in Shandong Agricultural University. China's big data began to shift from the construction phase to the application stage, and play an important role in the government decision making and social development.

3. CONCLUSIONS

As an important domain in data science and big data, remote sensing research and applications in agriculture have substantial development in China. Great progresses have been made in the basic research such as radiation transfer model development and quantitative inversion of key agricultural parameters. Remote sensing applications in cropland classification and crop mapping, crop growth monitoring, and crop yield estimation are widely established. The operational China's agriculture remote sensing monitoring system provides voluminous tenable information services in management and practices in agriculture.

Data science is an essential part in scientific research programs in China. Various scientific databases have been set up, including those in different ministries and at the national level. Data science and big data applications are not only for governmental agencies, but also for the many companies that are engaged in enhancing the production efficiency and competitiveness.

ACKNOWLEDGMENTS

This work was supported by the National Science Foundation of China (grant No. 61661136006), China Ministry of Agriculture "Introduction of International Advanced Agricultural Science and Technology Program (948 Program)" project (No. 2016-X38), and National Outstanding Agricultural Researcher Fund.

REFERENCES

Chen, S., Sun, J., 1997. Key technical links and solution ways of setting up a working system for yield estimations of the main crops of China by satellite remote sensing. Journal of Natural Resources 12, 363–369.

Chen, Z., Li, H., Zhou, Q., Yang, G., Liu, J., 2000. Sampling and scaling scheme for monitoring the change of winter wheat acreage in China. Transactions of the Chinese Society of Agricultural Engineering 16, 126–129.

Chen, Z., Zhou, Q., Liu, J., Wang, L., Ren, J., Huang, Q., Deng, H., Zhang, L., Li, D., 2011. Charms-China agricultural remote sensing monitoring system. In: 2011 IEEE International Geoscience and Remote Sensing Symposium (IGARSS), pp. 3530–3533.

Du, W., 2008. Scientific implementation of the golden farm project. Information China 1, 80–83.

Fang, H., Liang, S., Hoogenboom, G., Teasdale, J., Cavigelli, M., 2008. Corn-yield estimation through assimilation of remotely sensed data into the CSM-CERES-Maize model. International Journal of Remote Sensing 29, 3011–3032. http://dx.doi.org/10.1080/01431160701408386.

Feng, W., Yao, X., Zhu, Y., Tian, Y.C., Cao, W.X., 2008. Monitoring leaf nitrogen concentration by hyperspectral remote sensing in wheat. Journal of Triticeae Crops 28, 851–860.

Gantz, J., Reinsel, D., 2012. The digital universe in 2020: big data, bigger digital shadows, and biggest growth in the far east. IDC iView: IDC Analyze the future 2007, 1–16.

Han, J., Cercone, N., 1999. Dviz: a system for visualizing data mining. In: Pacific-Asia Conference on Knowledge Discovery and Data Mining. Springer Berlin Heidelberg, pp. 390–399.

He, Y., Pan, X., Ma, S., 2013. Estimation of LAI and yield of sugarcane based on SPOT remote sensing data. Transactions of the Chinese Society of Agricultural Machinery 44, 226–231.

Huang, Q., Li, D.D., Chen, Z., Liu, J., Wang, L.-M., 2012. Monitoring of planting area and growth condition of winter wheat in China based on MODIS data. Transactions of the Chinese Society of Agricultural Machinery 7, 31.

Jia, Y., Li, B., Cheng, Y., Liu, T., Guo, Y., Wu, X., Wang, L., 2015. Comparison between GF-1 images and Landsat-8 images in monitoring maize LAI. Transactions of the Chinese Society of Agricultural Engineering 31, 173–179.

Jiang, X., Xu, Z., Lou, J., 2001. Study on monitoring winter wheat sown areas in northern Anhui province by use of the CBERS-1 satellite remote sensing image. Geology Anhui 11, 297–302.

Jiang, Z., Chen, Z., Ren, J., Zhou, Q., 2012. Estimation of crop yield using CERES-Wheat model based on particle filter data assimilation method. Transactions of the Chinese Society of Agricultural Engineering 28, 138–146.

Ju, W., Sun, H., 1997. Estimation of flooded area with weather satellite remote sensing technique. Scientia Meteorologica Sinica 17, 131–142.

Liu, J., 2005. On the operational mechanism of the basic condition platform of science and technology. China Science and Technology Forum 56–59.

Liu, Z., Zhang, Q., 2014. Research overview of large data technologies. Journal of Zhejiang University Science 48, 957–972.

Ma, Y., Wang, S., Ke, L., Hou, Y., 2004. A preliminary study on the re-initialization/re-parameterization of a crop model based on remote sensing data. Acta Phytoecology Sinica 29, 918–926.

Niu, Z., Chen, Y., Sui, H., Zhang, Q., Zhao, C., 2000. Mechanism analysis of leaf biochemical concentration by high spectral remote sensing. Journal of Remote Sensing 4, 125–130.

National Science & Technology Infrastructure, National Earth System Science Data Sharing Infrastructure, 2014. Infrastructure Information. Available from: http://www.geodata.cn/.

Shi, Y., Ji, S., Shao, X., Tang, H., Wu, W., Yang, P., Zhang, Y., Ryosuke, S., 2014. Framework of SAGI agriculture remote sensing and its perspectives in supporting national food security. Journal of Integrative Agriculture 13, 1443–1450.

Song, C., You, S., Liu, H., Ke, L., Zhong, X., 2011. The spatial pattern of soil moisture in northern Tibet based on TVDI method. Progress in Geography 30, 569–576.

Sun, Z., Du, K., Zheng, F., Yi, S., 2013. Research and application prospect of big data in intelligent agriculture. Journal of Agricultural Science and Technology 15, 63–71.

Tian, Q., Gong, P., Zhao, C., Guo, X., 2000. A feasibility study on diagnosing wheat water status using spectral reflectance. Chinese Science Bulletin 46, 666–669. http://dx.doi.org/10.1007/BF03182831.

Wang, X., 2009. The progress, current situation and development trend of e-government in China. E-Government 44–68.

Wang, K., Zhou, Z., Liao, J., Fu, Y., 2015. Model for estimating tobacco leaf area index in Guizhou Karst mountainous areas based on SAR data. Acta Tabacaria Sinica 21, 34–39.

Wen, F., 2013. Agricultural big data and new opportunities for development. China Rural Science & Technology 10, 14.

Wu, J., 1995. The development of database in China. Network & Information 9, 11–15.

Wu, Y., 2004. The trend of science and technology innovation and the construction of the national science and technology fundamental condition platform. Research Dialectics Nature 20, 73–76.

Wu, B.-F., Zhang, F., Liu, C.-L., Zhang, L., Luo, Z.-M., 2004. An integrated method for crop condition monitoring. Journal of Remote Sensing 6, 1.

Wu, B.-F., Meng, J., Li, Q., Zhang, F., Du, X., Yan, N., 2010. Latest development of "CropWatch"—an global crop monitoring system with remote sensing. Advances in Earth Science 25, 1013–1022.

Wu, B., Meng, J., Li, Q., Yan, N., Du, X., Zhang, M., 2014. Remote sensing-based global crop monitoring: experiences with China's CropWatch system. International Journal of Digital Earth 7, 113–137. http://dx.doi.org/10.1080/17538947.2013.821185.

Xiao, X., Boles, S., Liu, J., Zhuang, D., Frolking, S., Li, C., Salas, W., Moore III, B., 2005. Mapping paddy rice agriculture in southern China using multi-temporal MODIS images. Remote Sensing of Environment 95, 480–492. http://dx.doi.org/10.1016/j.rse.2004.12.009.

Xu, G., 2003. Implementing scientific data sharing and enhancing national science and technology competitiveness. Liaoning Science and Technology 24–26.

Xue, L., Yang, L., Fan, X., 2005. Estimation of nitrogen content and C/N in rice leaves and plant with canopy reflectance spectra. Zuo Wu Xue Bao 32, 430–435.

Yan, F., Li, M.-S., Wang, Y.-J., Qin, Z.-H., 2006. Application of remote sensing technique to monitor agricultural disasters. Ziran Zaihai Xuebao (Journal of Natural Disasters) 15, 131–136.

Yang, B., 2013. Application of remote sensing technology in agricultural disaster monitoring of reclamation area. Modern Agriculture 66–67.

Yang, B.J., Pei, Z.Y., 1999. Definition of crop condition and crop monitoring using remote sensing. Transactions of the Chinese Society of Agricultural Engineering 15, 214–218.

Yang, L., Yang, Y., 2016. The spatial and temporal pattern of soil moisture in the west Liaohe river basin based on TVDI method and its influencing factors. Journal of Arid Land Resources Environmental 76–81.

Yang, G., Liu, Q., Du, Y., Shi, Y., Feng, H., Wang, J., 2013. Review of optical remote sensing imaging simulation of farmland radiation transfer process. Acta Scientiarum Naturalium Universityrsitatis Pekinensis 49, 537–544.

Zhu, Y., Sun, J., Liao, S., Yang, Y., Zhu, P., Wang, J., Feng, M., Song, J., Du, J., 2010. Research and practice of earth system science data sharing. Journal of Geographical – Information Science 12, 1–8.

Zhu, J., Shi, Q., Chen, F., Shi, X., Dong, Z., Qin, Q., 2016. Research status and development trend of remote sensing large data. Journal of Image and Graphics 21, 1425–1439.

Chapter 8

Data Visualization of Complex Information Through Mind Mapping in Spain and the European Union

Jose M. Guerrero
Infoseg, S.A., Barcelona, Spain

The soul never thinks without a picture.

Aristotle

1. DATA SCIENCE ECOSYSTEM IN THE EUROPEAN UNION

The First Report of the European Union Data Market Study (EU Data Market SMART, 2013) estimates that the number of data users will reach more than 1.3 million in 2020, whereas the overall data market may reach €111 billion.

In 2014, the European data industry had about 243,000 companies and 1.7 million in the information and communication technology (ICT) and professional services sectors. But there is a clear skills gap in big data and data science. There is a need for addressing this skills gap by bringing together researchers and industrial companies.

The next subsections describe the most important initiatives and organizations related to data science in the European Union (EU).

1.1 Horizon 2020 (Horizon 2020 Documents, 2016)

Horizon 2020 (H2020) is the biggest EU Research and Innovation program ever with nearly €80 billion of funding available over 7 years (2014–20). The goal is to ensure Europe produces world-class science, removes barriers to innovation, and makes it easier for the public and private sectors to work together in delivering innovation. Big data has an important role to play in H2020. The EU has a very clear strategy on the data-driven economy as an ecosystem of

Federal Data Science. http://dx.doi.org/10.1016/B978-0-12-812443-7.00008-9

different types of players interacting in a Digital Single Market, leading to more business opportunities and an increased availability of knowledge and capital. Data Science in general has a very important role to play. H2020 provides funding for research and innovation activities in the field of big data and open data. The communication "Connectivity for a European Gigabit Society" (European Commission, 2016b) recognizes connectivity as an essential enabler of the data economy. The Commission has proposed a set of measures to ensure everyone in the EU will have the best possible internet connection, so they can participate fully in the digital economy. Open data is a very important field for the commission, and in particular, European legislation on reuse of public sector information, nonlegislative measures supporting the opening up of public sector information and open data portals.

1.2 The European Data Landscape Monitoring Tool (European Data Landscape Monitoring Tool Documents, 2017)

The European Data Market study (DATALANDSCAPE, 2017) aims to define, assess, and measure the European data economy, supporting the achievement of the Data Value Chain policy of the European Commission. This strategy is focused on developing a data ecosystem of stakeholders driving the growth of this market in Europe. The study also plans to support the development of the community of relevant stakeholders in the EU.

Fig. 8.1 is a screenshot of an excerpt of the European data landscape monitoring tool for each of the EU28 member states. The user can select the reference year, the indicator of reference and one or more EU member states to discover the corresponding values.

The indicators available are data market value, data market share, data companies, data users, share of data companies, share of data users, data companies revenues, and impacts of the data economy. The key indicators of the data-driven economy are segmented by twelve industries: construction, education, finance, health, information & communication, mining, manufacturing, professional services, public administration, transport, utilities, retail, and wholesale.

1.3 Open Data Incubator Europe

The Open Data Incubator for Europe (ODINE) (Open Data Incubator for Europe Documents, 2017) is a 6-month incubator for open data entrepreneurs across Europe. The program is funded with a €7.8m grant from H2020 and is delivered by the seven following partners: University of Southampton (University of Southampton Documents, 2017), Open Data Institute (ODI) (Open Data Institute Documents, 2017), The Guardian (The Guardian Documents, 2017), Telefonica Open Future (Telefonica Open Future_Documents, 2017), Fraunhofer Institute (Fraunhofer Institute Documents, 2017), Open Knowledge Foundation (Open Knowledge Foundation Documents, 2017), and Telefonica (Telefonica Documents, 2017).

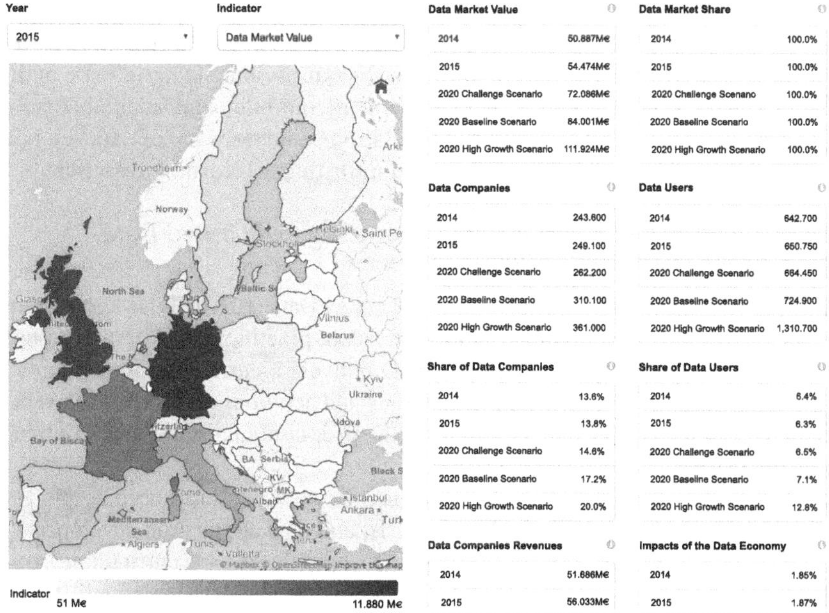

FIGURE 8.1 European data landscape monitoring tool.

ODINE aims to support the next generation of digital businesses and supports them to fast track the development of their products. The ODINE is already championing the best of European digital talent and looks forward to more outstanding companies to incubate.

1.4 Data Science Education in the European Union

In this section are described the most important institutions offering courses related to data science education in the EU.

1.4.1 European Data Science Academy

The European Data Science Academy (EDSA) (European Data Science Academy Documents, 2016) designs curricula for data science training and data science education across the EU. It is an initiative to fill the ICT skill gaps across Europe. It is publicly funded through the H2020 program. The leader of this project is the UK's Open University. The other organizations involved are the University of Southampton, Technische Universiteit Eindhoven (TH/e), Sweden's Kungliga Tekniska Högskolan (Kungliga Tekniska Högskolan Documents, 2017), Slovenia's Jožef Stefan Institute (Jožef Stefan Institute Documents, 2017), the ODI, Germany's Fraunhofer IAIS research body (Fraunhofer IAIS Documents, 2017), the UK's Persontyle Data Science Center of Excellence (Persontyle Documents, 2017), and France's ideXlab (ideXlab Documents, 2017).

Many of the learning materials developed by the EDSA are in multiple languages instead of only in English. EDSA tasks are as follows: analysis of the required sector specific skill sets for data scientists across the main industrial sectors in Europe, development of modular and adaptable data science curricula to meet industry expectations, and delivery of data science training supported by multiplatform and multilingual learning resources.

1.4.2 Educational Curriculum for the Usage of Linked Data (EUCLID Documents, 2017)

Educational Curriculum for the Usage of Linked Data (EUCLID) is a European project facilitating professional training for data practitioners, who aim to use Linked Data in their daily work. EUCLID delivers a curriculum implemented as a combination of living learning materials and activities (eBook series, webinars, face-to-face training), validated by the user community through continuous feedback.

Learning pathways at introductory, intermediate, and advanced levels for data architects, data managers, data analysts, and data application developers.

EUCLID wants to provide a comprehensive educational curriculum, supported by multimodal learning materials and highly visible e-learning distribution channels, tailored to the real needs of data practitioners.

1.4.3 Master Course in Data Mining and Knowledge Management

This master is offered by a consortium of five universities from three countries: **France** (Université Pierre and Marie Curie Paris 6 (Université Pierre and Marie Curie Paris 6 Documents, 2017), University of Lyon Lumière Lyon 2 (Université de Lyon Lumière Lyon 2 Documents, 2017), Polytech'Nantes (Polytech'Nantes Documents, 2017)), **Italy** (Università degli Studi del Piemonte Orientale (Università degli Studi del Piemonte Orientale Documents, 2017)), and **Spain** (Universitat Politècnica de Catalunya (UPC) (Universitat Politècnica de Catalunya Documents, 2017)).

This master course tries to face some important challenges. The first one is the exponential increase of data that cannot be solved by using larger storing devices or faster computers. New intelligent approaches are to be designed to tame the very size of data. Another challenge is that data can be in the form of numbers, texts, audio–video, sensor signals, and so on. Integrating into a unique system such complex data is still a challenge. Also, spatial distribution of data (for instance, on several websites or different databases) is a source of difficulty for integration.

The main reasons that have motivated the creation of this master course are the need for new skills and knowledge, the absence of specialized and integrated training, the pedagogical know-how of the partners and their scientific expertise.

1.4.4 European Institute of Innovation & Technology (EIT Documents, 2017) Digital Master School. Data Science Master (EIT Data Science Master, 2017)

Students can study data science, innovation, and entrepreneurship at leading European universities (Universidad Politécnica de Madrid (UPM) (Universidad Politécnica de Madrid Documents, 2017), TH/e (Technische Universiteit Eindhoven Documents, 2017), Université Nice Sophia Antipolis (UNS) (Université Nice Sophia Antipolis Documents, 2017), Politecnico di Milano (PUM) (Politecnico di Milano Documents, 2017), Technische Universität Berlin (TUB) (Technische Universität Berlin Documents, 2017), and Aalto University (AU) (Aalto University Documents, 2017)). This program includes scalable data collection techniques, data analysis methods, and a suite of tools and technologies that address data capture, processing, storage, transfer, analysis, and visualization, and related concepts.

The first year is very similar at all entry point universities (UPM, TH/e, UNS, and PUM). The second year enables students to concentrate on one of five technical specialization areas: Infrastructures for Large-Scale Data Management and Analysis (UPM), Multimedia and Web Science for Big Data (UNS), Business Process Intelligence (TU/e), Distributed Systems and Data Mining for Really Big Data (KTH), Design, Implementation, and Usage of Data Science Instruments (TUB) and Machine Learning, Big Data Management, and Business Analytics (AU).

1.4.5 Erasmus Mundus Joint Master Degree Program in Big Data Management and Analytics (BDMA, 2017)

The program favors the integration of students into a network of specialists and researchers in business intelligence (BI) and big data (BD). The curriculum is jointly delivered by Université Libre de Bruxelles (ULB) (Université Libre de Bruxelles Documents, 2017) in Belgium, UPC in Spain, TUB in Germany, TH/e in the Netherlands, and Université François Rabelais Tours (UFRT) (Université François Rabelais Tours Documents, 2017) in France.

Scholars from academic partners around the world and partners from leading industries in BI, private R&D companies, service companies, public research institutes, and public authorities contribute to the program by training students, providing computers, software, course material, job placement or internship perspectives, and financial support. The consortium prepares the students to answer today's professional challenges by a strong connection with the needs coming from the industry and also to pursue their studies into doctorate programs, through strong connections with the researchers' and innovators' views.

The first year is devoted to fundamentals: the first semester at ULB, the second one at UPC. Then, all students participate to the European Business Intelligence Summer School. In the third semester, the program offers three specializations: Large-Scale Data Analytics, Business Process Analytics, and

Content and Usage Analytics at TUB, TU/e, and UFRT, respectively, and students move to the partner of the chosen specialization. The fourth semester is dedicated to the master's thesis and can be realized at either as a placement in industry or as an internship in a research laboratory in any full or associated partner.

1.5 Other Organizations

1.5.1 Open Data Institute

The ODI helps commercial and noncommercial organizations and governments identify and address how the web of data will impact them. The ODI is an independent, nonprofit, nonpartisan company. They connect, equip, and inspire people to innovate with data by:

- Providing leadership and helping to develop strategy. Help to codevelop a clear strategic vision and focusing on timely, scalable, and relevant issues.
- Researching and innovating. Research user needs and business models, then develop and deploy open standards, licensing, data processing tools, and technology processes and techniques to meet these needs.
- Developing language and shaping policy. Define and debate issues around privacy, licensing, liability, risk and compliance, and related challenges regarding data infrastructure, data ethics, data equality, and the web of data.
- Giving training. Offer public and bespoke courses and online learning for data leaders, managers, and users.
- Supporting and encouraging start-ups. Offer a start-up program to create new jobs, products, and businesses and to drive investment and partnerships in data.
- Creating global networks. Connect people through events, media, and membership to discuss or work through organizational, local, or global challenges with data.
- Bringing the voice of business to UK government. Promote data innovation across government.

1.5.2 European Data Forum (EDF Documents, 2016)

European Data Forum (EDF) is a meeting place for industry, research, policymakers, and community initiatives to discuss the challenges of big data and the emerging Data Economy and to develop suitable action plans for addressing these challenges. The focus is not Small and Medium-sized Enterprises (SMEs) because they are driving innovation and competition in many data-driven economic sectors. The most important topics discussed in the EDF are novel data-driven business models, technological innovations, and societal aspects.

The EDF series is a yearly event on social, economic, research, engineering, and scientific topics centered on the arising European data-driven economy. Each event defines a specific focus and theme. These are defined in a mission report written by the event chairs and approved by the steering committee. A major focus of the event series is to raise issues that are of interest to European SMEs; however, EDF also aims to provide a meeting place where they network with colleagues from research, policy making, societal initiatives, and from large industry.

EDF events are envisioned to run under the umbrella of the respective EU presidency.

1.5.3 Big Data Value Association (BDV Documents, 2017)

The Big Data Value (BDV) is a fully self-financed industry-led association representing the BDV stakeholder community with a presence in Europe. The objectives of the association are to boost European BDV research, development, and innovation and to foster a positive perception of BDV. They are interested in research and innovation in data management, data processing, data analytics, data protection, and data visualization. The objectives of BDV are to boost European BDV research, development, and innovation and to foster a positive perception of BDV. Between the founding members we can find IBM, Nokia, Intel, SAP, Insight, Siemens, SINTEF, and Atos. The BDV cooperates with the European Commission. BDV association membership is composed of large industries, SMEs, and research organizations to support the EU BDV public–private partnership.

1.6 Data Science and Big Data in Spain

In Spain the most interesting learning initiatives, besides the cited above at the EU level, are as follows:

- Master in Data Science (Universidad Juan Carlos I, 2016). Universidad Juan Carlos I (Universidad Juan Carlos I Documents, 2017).
- Master in Business Intelligence and Big Data. Universitat Oberta de Catalunya (Universitat Oberta de Catalunya Documents, 2017).
- Master in Business Analytics and Big Data. Instituto de Empresa (Instituto de Empresa Documents, 2017).
- Master's Degree in Data Science. Barcelona Graduate School of Economics (Barcelona Graduate School of Economics Documents, 2017).
- Master Telefonica in Big Data (Telefonica, 2016).
- Master Foundations of Data Science (Universitat de Barcelona Documents, 2017).
- MSc in Business Analytics. ESADE Business School (ESADE Documents, 2017).

Some of the most active companies and centers are:

- Infoseg (Infoseg Documents, 2017). Specialized in the development of software for the visualization of Open Data, Big Data, and any type

of complex information using mind mapping automation. They have created applications in the healthcare, insurance, banking, information security, geographic information system (GIS), manufacturing, customer relationship management (CRM), content management system (CMS), and artificial intelligence (AI) fields.
- Barcelona Supercomputing Center (BSC) (BSC Documents, 2017).
- Synergic Partners (Group Telefonica) (Synergic Partners Documents, 2017). They organize the data science awards.
- Big Data Spain (Big Data Spain Documents, 2017). They organize an important annual event in Big Data.
- Piperlab (Piperlab Documents, 2017).
- BCN Analytics (BCN Analytics Documents, 2017).
- Geographica (Geographica Documents, 2017). GIS and Data Science for companies and smart cities.
- Innaxis (Innaxis Documents, 2017). Data Science. It is a private, nonprofit research institute that addresses complex scientific challenges of significant social and economic impact in different sectors: aviation, transport, ICT, life sciences, and social sciences.
- BIGDATACoE (BIGDATACoE Documents, 2017). Center of Excellence of Big Data in Barcelona. Dedicated to build, develop and provide tools, data sets and value-added Big Data capabilities to enable companies the definition, testing, and validation of Big Data models before its final implementation
- Bismart (Bismart Documents, 2017).

2. OPEN DATA IN THE EUROPEAN UNION AND SPAIN

In the EU, 27 countries have a national open data portal. A new study (European Data Portal, 2015) has been published to assess the open data maturity status of the EU state members together with Liechtenstein, Norway, and Switzerland (UE28+).

2.1 Open Data in the European Union

Open Data in the EU is centralized in the data.europe.eu website (data.europa.eu Documents, 2017).

There, three types of information can be found: The EU open data portal (EU Open Data Portal Documents, 2017), European Data Portal (European Data Portal Documents, 2017), and resources with persistent URIs of the EU institutions and bodies.

The EU Open Data Portal is the single point of access to open data produced by EU institutions and bodies.

All data contained in this site are free to use, reuse, link, and redistribute for commercial or noncommercial purposes.

This portal is divided in four areas:

- **Data**, as data sets that can be viewed or downloaded in HTML, PDF, ZIP, Text/tab-separated values, RSS, and XML formats.
- **Applications** developed by the European institutions, agencies, and other bodies as well as third parties.
- **Linked Data**, makes it easier for developers to connect information from different sources.
- **Developer's Corner**. Two programming interfaces are offered to developers, the REST API and the SPARQL Endpoint.

The European Data Portal contains the metadata of public sector information available on public data portals across European countries.

2.2 Open Data in Spain

In Spain there three levels of open data websites: national, autonomous communities, and city councils.

At the national level we find the datos.gob.es website (datos.gob.es Documents, 2017).

Here there are three types of users of the data sets: citizens that need the information, public organisms that provide or consume the information, and reusers that create products or services that use the information provided.

At the autonomous community level a good example is the one of Catalonia (Open Data gencat Documents, 2017). At the city council level, one of the best examples is Barcelona (OpenDataBCN Documents, 2017).

Many cities are beginning to have citizen's file systems containing a record of all the information of each citizen, including tax and census, as well as the progress of their official business. This information is open to the citizens who remotely identify themselves in a secure way through a digital identity (Barcelona Carpeta del Ciudadano Documents, 2017).

3. VISUALIZATION OF BIG DATA AND OPEN DATA

There are two traditional ways of presenting complex information contained in Big Data and Open Data websites: linear text (or linear presentation of several elements of information), and websites (or classical software applications using hyperlinks).

The main problems of linear text are given as follows(Guerrero, 2016): it does not provide a sense of perspective or structure; it does not display the relation between the parts (Agichtein and Gravano, 2000); it makes impossible to see the "whole picture"; it is difficult to understand when it is long; it is impossible to use for collaboration; the more complex the information, the more difficult it is to assimilate it; when a linear text document has to be reviewed, a lot of time is needed to analyze the information again.

Websites and traditional software applications containing hyperlinks have basically the same problems as linear text plus the added complication of disorientation and loss of context when navigating (Ahuja and Webster, 2001). This was soon detected in the first years of the World Wide Web, and the system of navigation using "breadcrumbs" was created to limit the extension of the problem (Fillion and Boyle, 1991). But this does not solve the issue completely. It only makes the situation a little better.

In the case of Big Data, we try to get insight into data sets and other sources of information by using analytics. The result is usually a large number of tables and charts that present similar problems to linear text. When trying to organize them by using web pages and hyperlinks, problems arise because the human working memory is limited to 3–5 elements of information (Cowan, 2010). Irrelevant information occupies space in our working memory. We have problems to filter irrelevant information. Each change from a web page to a new one consumes even more of the limited resources of the working memory.

In the case of Open Data and depending on the subject, information can usually be downloaded from websites in the following formats: PDF, PNG, Excel, HTML, XML, ZIP, RSS, Text/Tab-separated-values, CSV, Access, PPTX, DOC. When several files are downloaded, the user has to document and relate them between themselves. This can be a serious problem if the number of files is large (more than five or six). The content is not indexed, and there are problems to find the information that is required.

The only apparent solution is to create mini websites containing the information created, in the case of big data analytics, or the different files downloaded, in the case of open data, plus complementary information documenting the content of each data set or file. This solution is unfeasible for most users. The users have to keep downloading, opening, and trying to understand what is contained in each individual file.

4. MIND MAPPING

4.1 Introduction

Mind mapping is a graphical technique for visually displaying and organizing several items of information. Each item of information is written down and then linked by lines to the other pieces thus creating a network of relationships. They are always organized around a single central idea (Guerrero and Ramos, 2015).

A mind map is a diagram created by mind mapping.

Fig. 8.2 shows an example of a simple mind map corresponding to the interactions of aspirin.

A summary of the history of mind mapping can be found in (Guerrero and Ramos, 2015; Guerrero, 2016).

The most important reasons why we prefer visual information to simple text are (Ware, 2004) defined as follows: visual information provides an ability to comprehend large amounts of data, visual information allows the perception

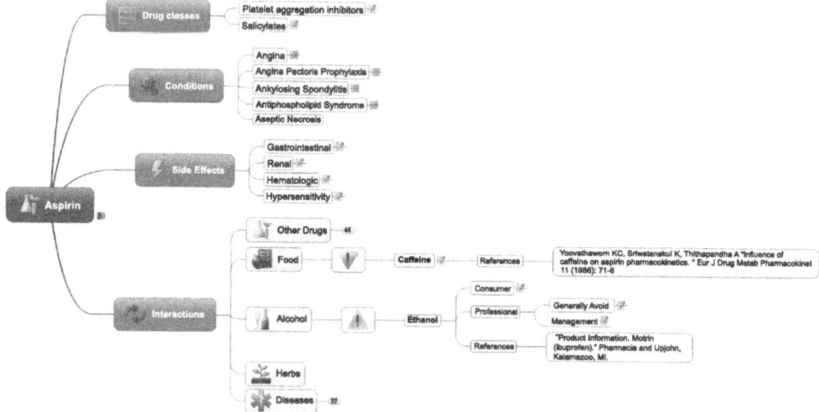

FIGURE 8.2 Interactions of aspirin.

of emergent properties that were not anticipated, visual information enables problems with the data itself to become more immediately apparent, visual information facilitates understanding of both large-scale and small-scale features of the data, and that the use of visual information facilitates the formation of hypothesis.

Mind mapping solves the above mentioned problems of linear text because its content is easy to analyze, understand, and memorize; it provides a sense of structure and perspective because of the visual display of the information; and mind mapping makes it easy to see the relation between the parts of the information. But probably the greatest advantage of mind maps is that they make it very easy to see the "whole picture" and the details at the same time.

In general, mind maps can be displayed on a single screen or page, eliminating in this way the problems related to the use of hyperlinks. This reduces in a significant way the cognitive load of the user when viewing and analyzing the information displayed on the mind map. Human beings do not get disoriented or lost while navigating mind maps (Guerrero, 2016).

4.2 Digital Mind Maps

A digital mind map is a single compressed file that can be viewed using mind mapping software. The most common elements of a digital mind map are given below:

- Topic: node containing a short text and an icon or image. This is the fundamental element of mind maps.
- Line: between two topics to represent their hierarchical relationship.
- Note: long text attached to a topic. Used to contain long pieces of complex text and images.

- Hyperlink: link to a web page or another mind map. It is very useful to help to organize information.
- Callout: visual comment on a topic to focus the attention of the viewer.
- Relationship: link between two topics. Different from the hierarchical relationship described with the lines. It is usually nonhierarchical.
- Boundary: shape surrounding a topic and all its subtopics. This helps to group all the elements of a branch of the tree.
- Spreadsheet: contained in a topic. Tables to organize numerical information.
- Chart: contained in a topic. Simple graphics to describe visually numerical information.

These elements are the most common in all mind mapping software products. However, some of them such as spreadsheets, charts, callouts, or hyperlinks are missing in many of them. This is an important factor to take into account at the time of choosing a mind mapping software application. MindManager (MindManager Documents, 2017) is an example of application that has the possibility of using all of the above elements. This is one of the reasons why MindManager is the most used software in mind mapping for professional application. The available environments for MindManager are as follows: Windows, Mac, iPad, iPhone, Android, and Web.

MindManager is used by 83% of all companies in the Fortune 100 list (Mindjet Documents, 2017).

A good example of big companies using mind mapping are Boeing (Mind mapping at Boeing: Getting off the ground, 2016) and its European competitor Airbus (Ben Hamida et al., 2016).

4.3 The Importance of Mind Mapping

Mind mapping is needed because we have serious problems of information overload (Toffler, 1970), excess of linear text, excess of complex information in websites, and traditional software applications using hyperlinks.

An increasing amount of information is being produced and we cannot cope with it. The widespread access to the Web and email has increased the magnitude of the problem (Hemp, 2009). Mind mapping helps to organize and understand the information much more effectively than any other technique, reducing the problems created by information overload.

At the same time, the complexity of the information we have to manage is also growing. This fact makes it even more complicated to try to manage the information overload.

Attention is needed to analyze all the information we receive. We need working memory to control our attention (Corbetta and Shulman, 2002). The brain cannot multitask when it comes to paying attention.

When reading linear text we can only use controlled attention. With mind mapping we can use controlled attention and stimulus-driven attention due to the visual elements that it includes.

The excessive use of linear text makes it more difficult to analyze and understand information. The human brain is not prepared to work with text. The first written language is barely 5000 years old. Our brain has not had time to evolve and improve our text processing capabilities. The visual part of our brain has had millions of years to evolve and optimize our visual capabilities (Lyon, 2007).

Generic websites, open data websites, big data analytics websites, and traditional software applications contain complex information that is as difficult to understand as linear text or hypertext.

We try to solve these problems by using continuous partial attention (Stone, 2006). This is paying partial attention continuously. We try not to miss anything. But this strategy cannot be successful because of the limitations of our working memory.

4.4 Advantages of Mind Mapping

Mind mapping does not completely solve all serious problems related to information overload and the complexity of that information, but it is the best tool we have to minimize the negative effect of those problems.

Mind maps allow the use of the right mix of visuals and text. The "whole view" and the details can be seen at the same time. Everything can be seen on the same screen (without screen changes). Mind mapping uses stimulus-driven attention to process visual information faster, reducing, or eliminating errors in understanding. Mind maps can be used by any type of user, without the need to be an expert. The integration of mind mapping software with multiple applications enables users to transform complex information into actionable project plans.

The main reasons why mind mapping is so effective are the fast processing of images in the brain (Nelson et al., 1976), the hierarchical organization of mind maps (Bower et al., 1969), the chunking of information in smaller parts (Miller, 1956), and the viewable relationship between the parts. Beside those, digital mind maps provide some extra advantages over linear text and web pages: the possibility of collapsing/expanding the branches; the integration of multimedia elements; mind maps are basically a single compressed file, easy to email or upload to websites using FTP; they facilitate online collaboration; the automatic self-reorganization after modification of the content.

In linear text, part of the information is only implicit and has to be computed to make it explicit. Usually, when working with complex information, the cost of making it explicit for use is very costly (Larkin and Simon, 1987). Mind maps, as diagrammatic representations, are indexed by location in a plane and have a lot less implicit information that needs to be made explicit. Mind maps are much more efficient in the search and recognition of information and in the process of making inferences.

This combination of features is unique in mind mapping and makes it an essential tool in the management of complex information.

Sooner or later all organizations reach a point where they need tools such as mind mapping to be able to see the whole picture of the complex information they have to deal with.

This is the case of the UK Government, which have launched a "Transformation Map Project" to "develop a dynamic mapping tool of government transformation strategy across government, highlighting dependencies and programme information" (GOV.UK Digital Marketplace, 2017). It is interesting to read the description they make of the problem to be solved: "Departments cannot currently see how their programme relates to other programmes across government. This in turn leads to activities being duplicated across government. Departments cannot see how other programmes depend on them, leading to poor decisions for the whole of government being taken. Ministers cannot see how their department has to work with other departments to achieve their priorities. This leads to resourcing being improperly allocated across the department. Treasury cannot plan cross government programmes as they cannot see how these programmes fit into wider government transformation." Difficulties in visualizing complex information lead to low productivity, poor decisions, and improper allocation of resources. In situations like these, mind mapping is probably the only possible solution, and organizations are realizing it.

4.5 Experiments and Surveys Related to Mind Mapping

There are many studies providing evidence of the advantages of using mind mapping. One of the most interesting is *The efficacy of the* "mind map" *study technique* (Farrand et al., 2002). They found that there was a significant performance difference on memory recall between a group studying using traditional note-taking and another group that employed mind mapping. Improvement was estimated in a 15%.

In another study (Holland et al., 2004), it was found that mind mapping clearly helped students in improving the structure, coherence, and quality of their written work.

But most of what we know about the advantages of the use of mind mapping have been obtained from surveys between users of mind mapping software. The most interesting survey is the one published by the Mind Mapping Software Blog, the *2015 Mind Mapping Software Trends Survey* (Frey, 2016) that focuses in the way in which business executives use mind mapping software in their work. In Fig. 8.3 are shown the results corresponding to the *Productivity increase experienced from using mind mapping*.

The average increase in productivity has been found to be 20%–30% consistently in four surveys during a period of 8 years. Interestingly, a percentage of about 3% reported increases in productivity of 100%. This usually corresponds to the answer of the most experienced users of the mind mapping technique.

To the question "To what extent does your mind mapping software help you to distill information and reach clarity faster?", more than half of the respondents answered that "significantly" and 25% that "it is essential."

Productivity increase experienced from using mind mapping (%)

2015 Mind Mapping Software Trends Survey (Frey 2016)

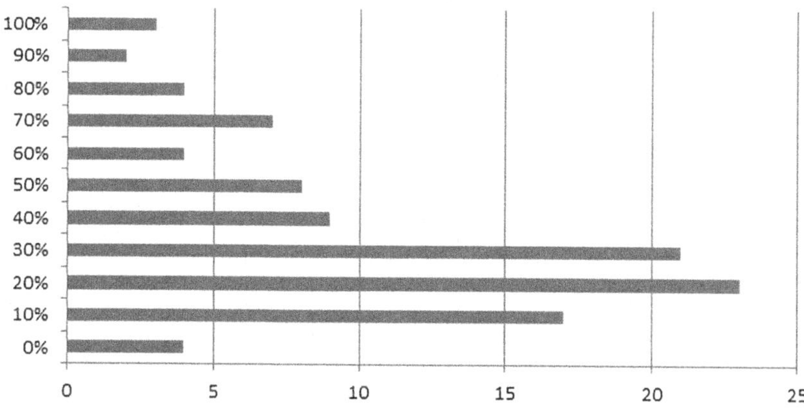

FIGURE 8.3 Productivity increase experienced from using mind mapping.

Main uses of mind mapping in business applications (%)

2015 Mind Mapping Software Trends Survey (Frey 2016)

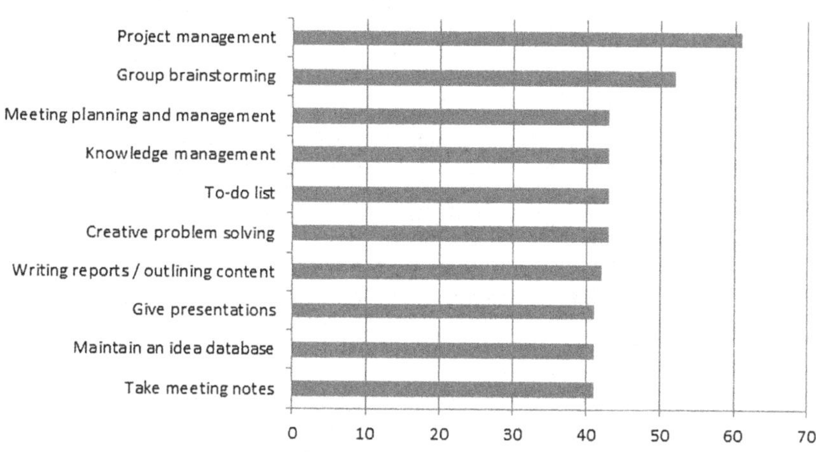

FIGURE 8.4 Main uses of mind mapping software.

To the question How much of a creativity increase have you experienced from using mind mapping software? 20% of the respondents answered that "30%" and 5% that "100%."

The main uses of mind mapping software in the business applications are the ones shown in Fig. 8.4.

The most important single benefits of the use of mind mapping are shown in Fig. 8.5.

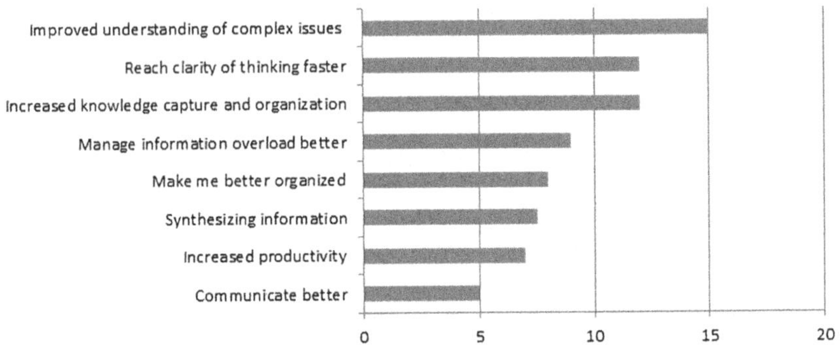

FIGURE 8.5 Single biggest benefit experienced from using mind mapping software.

4.6 Use of Mind Mapping in Governments

Use of mind mapping is not as widespread in governments as it is in the business and education fields, but it is beginning to have a significant presence. Some examples are given below:

Use of mind mapping in large departments of the Dutch government (Van Oudsheuden, 2013a,b). The mind mapping technique is used in impact analysis, in policy making to find solutions to complex problems, and in brainstorming and creativity.

Analysis of the infrastructure policy in the UK government (Insight Public Affairs, 2012). Mind mapping has been used to visualize the whole picture of a very complex infrastructure policy. A sample mind map looks at some of the headline policies, which the government has developed on key issues such as aviation, rail, energy, roads, and planning.

Visualization of legal information (Brunschwig, 2006). This article proposes the use of mind mapping to visualize legal information to improve the experience of e-government websites users.

Visualization of the impact of community factors in Australia (Department of Health, 2004). Mind mapping is used at key informant interviews, conversations, stakeholders' meetings, and focus groups.

Visualization of complex information in Australia (Department of Finance, 2013). Mind mapping is used to visualize the government data landscape.

Demand analysis in e-government in China (Wu et al., 2009). Mind mapping is adapted for demand analysis, offering several benefits to e-government planning.

Knowledge management and sharing in Taiwan (Ho and Lin, 2013). In a study, mind mapping was applied to enhance knowledge management performance in a construction project. The intention was to help engineers to reuse and share knowledge and experience. Mind mapping is an excellent tool for knowledge management and sharing because of its flexibility and multimedia elements.

Testing of website usability standards in New Zealand (Burton, 2013). Mind mapping software is used to provide a visual representation of usability testing in a lean testing initiative. This application is becoming very common in the business field.

Scenario planning in Vietnam (Nguyen et al., 2014). Mind mapping is used in scenario planning to solve complex sanitation problems. Group brainstorming was performed using an initial mind map.

Exploratory testing in the United Kingdom (GOV.UK Exploratory Testing, 2016). Mind mapping is cited, with text charters and time boxing, as one of the techniques to achieve best results during exploratory testing.

Defense project management in Australia (McConnachie, 2009). Mind mapping has been used in the project development of the F/A-18F Super Hornet Air Vehicle Weapon System Database, the development of the user help manuals and web help interface, project development planning for the ANZAC Ship Boat Crane training, integrated logistic support project management for the Ground Surveillance Radar Project for the Defense Materiel Organization and in development and process management for a review of the Defense Land Rover Fleet of Vehicles.

Rapid qualitative data analysis in public participation processes in the United Kingdom (Burgess-Allen and Owen-Smith, 2010). The article draws on experience of using mind maps in several participation processes, but a single example is used to illustrate the approach, patient participation focus groups. The use of mind mapping to manage qualitative data offers an interesting solution when there are limited resources in patient participation processes.

To understand the social situation in problematic neighborhoods in France (Mongin, 2005). For the diagnosis and understanding of the interactions between people and the administrations.

To visualize redacted proposals from the White House's open government brainstorming on transparency, participation and collaboration led by Office of Science and Technology Policy's open government team in the United States (DebateGraph Documents, 2017). The proposals are open to collaborative editing. People can also add supporting and opposing arguments to them.

The previous examples are mostly of manual mind mapping using software to edit and visualize mind maps, in a few cases even pen and paper mind mapping is still used.

4.7 Mind Mapping Automation

The creation of mind maps can be automated in several ways. One of them is by using the application programming interfaces (APIs) provided by the manufacturers of the different mind mapping software products.

The company Infoseg has created a new approach to the mind mapping automation. It is based on the use of an XML standard file format. Any information expressed in that format can be automatically converted into a MindManager file and sent as an email attachment of uploaded to a website using FTP. This technology reduces considerably the amount of time needed to create applications to visualize complex information or to organize that information to make it accessible to any type of user.

Information contained in databases, text files, or websites can be used to create XML files that are easily converted into a MindManager mind map using the software developed by Infoseg.

A detailed description of the process of dynamic mind mapping can be found in the Infoseg website (Infoseg, 2006).

In the following subsections, some of the most interesting examples of automation applications developed are described. The ones corresponding to Infoseg are SlideShare presentations.

4.7.1 Statistical Analysis With R (Infoseg Documents, 2013a,b)

This is a model of how information from statistical analysis should be presented. Even a very simple statistical analysis produces a large number of pages containing linear information and charts. A mind map offers the same information but better organized and in a format that is easier to understand by nonexperts. This is the basis for the visualization of big data analysis results. In this case R has been used as the tool for the statistical analysis.

4.7.2 Big Open Data Analysis Using R (Infoseg, 2014a,b)

This is an example of the visualization of the medicare provider utilization and payment data: physician and other supplier (CMS.gov, 2014). It is the information on services and procedures provided to medicare beneficiaries by physicians and other healthcare professionals. Contains information on utilization, payment (allowed and medicare payment), and submitted charges organized by National Provider Identifier.

4.7.3 Open Data (Infoseg, 2013c,d)

This SlideShare presentation offers an introduction to the problems of open data websites and the use of mind mapping as a solution. Other tools can also be used with mind mapping.

This is a generic application of mind mapping to organize information from a public website.

Another possible application is the visualization of information of clinical trials from open data portals. Clinical trials generate safety and efficacy data. The sponsor may be a governmental organization or a pharmaceutical, biotechnology, or medical device company.

The presentation clinical trials open data (Infoseg, 2014c,d) offers an example of a single study from the clinicaltrials.gov website (ClinicalTrials.gov Documents, 2017). It is a registry and results data base of public and private studies from around the world and that has results in XML format. These results can be downloaded and analyzed to generate mind maps containing the details of individual studies.

4.7.4 Complex Information Systems

Systems such as CMS, CRM, and others are very powerful but have the problem that they contain too much information that cannot be managed efficiently by using linear text or hypertext. In the case of Microsoft SharePoint (Microsoft Sharepoint Document, 2017), a popular CMS, the mind mapping software MindManager has been integrated into SharePoint as a front end to visualize the complex information it contains (MindManager Enterprise Documents, 2017).

The same has been recently done for Microsoft CRM (Microsoft CRM Documents, 2017) by Infoseg (Infoseg, 2016).

Other complex information systems such as SAP (SAP Documents, 2017) can also benefit from the advantages of mind mapping (Infoseg, 2014e).

4.7.5 Analysis of Information Security Log Files

With increasing frequency information security log files have to be analyzed. These log files are generated by desktop computers, servers, routers, switches, firewalls, proxy servers, virtual private networks, and other network systems. The results of these analyses are increasingly more complex and difficult to use to make decisions. Mind mapping is an excellent tool to visualize those complex results in a way that facilitates the work of the persons who have to make those decisions. Infoseg has created applications to visualize the results of the analyses as mind maps (Infoseg, 2014f).

4.7.6 Mind Mapping Integration in Artificial Intelligence Systems

AI systems such as IBM Watson (IBM Watson Documents, 2017) can analyze data, unstructured text, images, audio and video, producing insightful information. It does even offer cloud access to complex analytics and visualization (IBM Watson Analytics Documents, 2017) through a set of open APIs and SaaS products. However, these visualizations are in the form of dashboards that have some of the problems of linear text and software applications using hyperlinks.

Infoseg has started a project to use Watson's open APIs to try to solve those visualization problems by generating digital mind maps directly from Watson.

Infoseg started this project with Alchemy API (IBM Alchemy Documents, 2017), a set of 12 semantic text analysis APIs using natural language processing. Infoseg has developed an application in C#.NET that generates mind maps directly from Watson output (Infoseg, 2017). Now Infoseg is planning to develop an add-in for Microsoft Outlook that will allow users to analyze their email messages before sending them.

This is just the beginning of a huge set of applications that can be developed to integrate mind mapping into the IBM Watson environment. The technology developed by Infoseg allows the creation of new applications in days or weeks instead of in months or years, these estimations are supported by the experience of Infoseg in the development of this kind of applications in many fields.

In Fig. 8.6 we can see an example of a mind map generated by the software developed by Infoseg using the Alchemy API. It is an analysis of the article

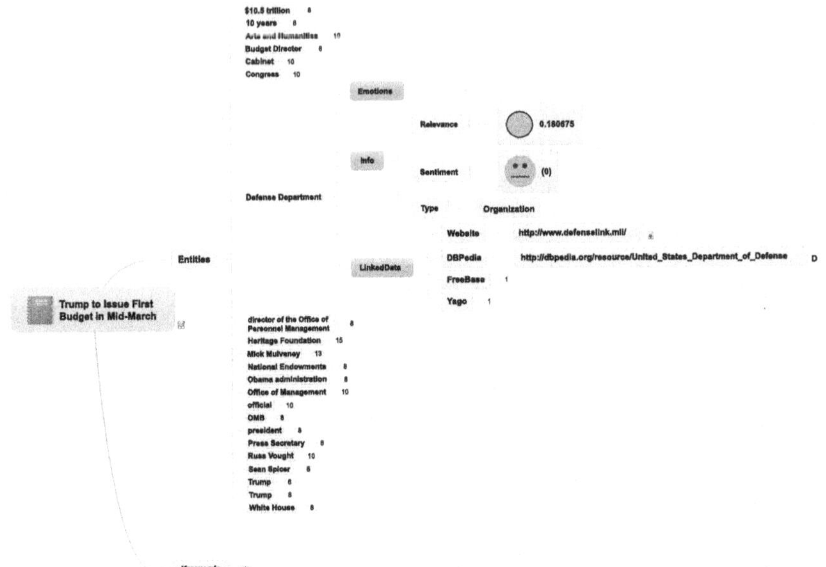

FIGURE 8.6 Integration of mind mapping automation into IBM Watson.

Trump to Issue First Budget in Mid-March, Promising "A Lot More With Less" (Govexec.com, 2017). It is a relatively simple analysis using just entities and keywords, but it is a useful example of the great possibilities of the use of mind mapping with cognitive computing.

Besides Alchemy, other IBM Watson products that can benefit from the use of mind mapping visualization are as follows: commerce, education, financial services, Internet of things, marketing, supply chain, talent, work, and health.

5. USES OF MIND MAPPING IN THE FEDERAL GOVERNMENT

The possible applications of mind mapping in the federal government can be classified into the following groups:

1. **Management Reporting**. Including key performance indicators to allow the directors and upper management to know how each element of the hierarchical federal structure is doing operationally. This should be an automatic process generating mind maps periodically at all levels of the structure: department, agency, division, service, and office. Some candidates are as follows: the whole Department of Agriculture, or any other, or a smaller unit like the United States Forest Service. But all departments or smaller units can benefit from the use of mind mapping for management reporting.
2. **Visual Project Management**. Mind mapping is probably the best way to manage complex projects visually. The number of projects is growing

continuously and with it the amount of money and resources dedicated to them. It is essential to optimize the use of those resources. A combination of mind mapping software and traditional project management software such as MS project is advisable. The mind map acts as a front end for the project management software. Complex projects simply cannot be understood without the use of visual techniques such as mind mapping.
Some candidates are as follows: the United States Forest Services, the Army Corps of Engineers or the Federal Energy Regulatory Commission, the US Department of Energy.

3. **Strategic Planning**. A good introduction to the use of mind mapping in strategic planning can be found in the article "6 Steps to Strategy Using Mind Mapping" (MacDonald, 2015). Some candidates are as follows: the Office of Policy and Strategic Planning of the Department of Commerce, the Office of Environment Safety and Health, or the Administration for Children and Families.
4. **Provision of advice to upper levels of the federal structure**. When the advice is complex, it is advisable to create the corresponding document using manual mind mapping. It facilitates the understanding of the advice and increases the efficiency of the process.
   Some candidates are as follows: the Office of Science and Technology, the Office of Management and Budget or the president's Advisory Committee on Trade Policy and Negotiations.
5. **Provision of information to upper levels of the federal government**. Again, when the information to be delivered is complex, it is advisable to use manual mind mapping to create an organized set including the basic information and all complementary information to document and clarify it. The mind maps can be created manually or automatically on a periodic basis.
   Some candidates are as follows: the Agricultural Marketing Service to provide information to the Price Support Division, the Office of Technology Utilization or the National Center for Health Statistics.
6. **On-demand provision of information to citizens and companies**. When the information requested by citizens and companies is complex, all agencies of the federal government should have the option of creating a mind map that would reduce the complexity of the answer. This option could be manual of automatic for certain cases. If the information requested is more complex than a simple chart, text page or spreadsheet, it will be worth analyzing the possibility of automating the answer to generate mind maps attached to email messages.
7. **Visualization of Big Data & Open Data**. These applications can be manual or automatic depending on the case. Mind mapping should be an option in any application involving the use of Big Data or Open Data. The most obvious candidates are as follows: the National Agricultural Statistics Service, the Economic Research Service, the National Resources Inventory,

or the Bureau of the Census because of the amount of public information they have to offer.

8. **Incident Reporting summaries**. Incident reporting systems are becoming commonplace (US-CERT, 2016). But no incident reporting system is complete without including the option of generating mind maps containing summaries of the most important incidents. These summaries can be generated automatically in a periodic way or when certain important incidents take place. Designated persons should have the means to receive emails on their tablets or smartphones with attached mind maps to facilitate the visualization of the most important incidents.
   All agencies should have a list of people who have to receive automatic email messages containing these emergency summary mind maps. The difference with simple SMS messages is that the mind maps can contain complex summaries well organized and with visual cues that facilitate the understanding of the most interesting information with a reduction in the possibilities of errors. Some clear candidates are as follows: Inspection of agricultural products for public safety, the risk management agency, the law enforcement agencies, the military. In these four cases incidents are very common, and incident reporting is a very useful tool.
9. **Organization of complex information**. When complex information has to be visualized frequently it is very useful to have the possibility of viewing it using mind mapping. In some cases that information will be already contained in any of the multiple databases of each agency of the federal government. Even in these cases it can be useful to have the option to create mind maps that can be manipulated manually to add new information not already defined in the tables of the data bases or to visualize the existing information in a more productive way without the use of the traditional web applications that use hyperlinks. All agencies should have this option to increase productivity and efficiency.
10. **Clerical work**. All agencies should offer their personnel the possibility of using mind mapping software to do manually their clerical work in a more efficient and productive way. One of the possible uses is that of organizing complex information to visualize and analyze it more effectively. The information can be about a person, a group of persons, a company, a group of companies, a geographical unit, an issue, a task, a problem, a case, a report, a book, an article, a web page or site, or a document. Another very important type of use is in the preparation of a report, an article, a book, a presentation, a class, a course, a meeting, or a briefing.
11. **Creativity**. Mind mapping helps to reorganize problems and view them from different perspectives and discover the real problem to solve. It does also help in brainstorming individual or group sessions (Ben Hamida et al., 2016).
12. **Education**. Working with mind maps helps students to organize ideas better to understand them more easily. It does also help when taking notes

in classes, presentations, and meetings. Teachers can use mind mapping to prepare classes and exams. For a more detailed description see (Guerrero, 2016).

13. **Knowledge and expertise management, sharing and transfer**. The features of mind maps make them a very good tool for this. In a single file, experts can store and exchange heterogeneous knowledge that can be examined and analyzed by nonexperts with only the help of free mind map viewers or even just browsers with HTML5 capabilities. The mind mapping software is very simple to use and complex information describing the knowledge of experts can be stored as short text, icons, images, notes, attachments, charts, spreadsheets, relations, and hyperlinks to web pages.
14. **Front end to visualize information contained in complex information systems**. Systems such as CMS, CRM, SAP, and any other system that works with complex information are very clear candidates for the use of mind mapping as a front end to facilitate the organization and information of this type of content.

The list of possible applications is very long and difficult to summarize in a single chapter. For example, just in the field of Law Enforcement, we could mention these applications: incident reporting, sharing of complex information, fight against cybercrime, CSI/forensics, analytics visualization, supervision and management, intelligence gathering and organization, interviewing, interrogation, case management.

6. CONCLUSIONS

The mind mapping technique has proved its efficacy in the business and education fields. Average productivity increases of around 30% have been reported in several surveys. There are no reasons why those increases cannot be achieved at the federal government level. An effort should be made to start pilot tests of manual and automation applications of mind mapping at all levels of the administration. The cost of introducing mind mapping software for manual and automated use is very reasonable and can be rapidly paid in full by the increase in productivity of millions of federal employees.

Every employee can learn to use mind mapping in a relatively short time, just a few days. Even the general public can receive and understand easily complex information that the federal government could create using mind mapping automation. The possibility of generating mind maps using HTML5 creates very interesting possibilities because citizens do not even need special viewers to work with mind maps. Any browser is now sufficient to visualize and navigate mind maps containing very complex information generated by the federal administration.

An easy way to evaluate the cost of not using mind mapping in the federal administration is by considering just a pessimistic increase in productivity of only a 10% by the use of this technique. To get an idea of the potential savings we just have to think in the 7,544,802 full-time federal employees in education with a payroll of $34,387,882,393 (US Census Bureau, 2015). Each of them is a very clear potential user of the mind mapping technique because of their type of work.

Many types of application for dealing with complex information can only be developed effectively by using mind mapping because linear text and hyperlinked text desktop applications, or web pages have too many problems because of the nature of the human brain.

The use of mind mapping to visualize complex information can also save a lot of time and effort because there is no need to develop such a big number of complex web pages. Websites can be simplified by using mind mapping automation to generate ad hoc information from those web pages. The user can visualize complex information in a more simple way and without the disorientation produced by web pages. Moreover, mind maps generated from websites can be downloaded into tablets and smartphones and accessed anywhere without the need or cost of access to the Internet. Some obvious examples are personal health records of passengers in cruise liners and travelers to foreign countries.

AI is making its way at all levels of the federal government. This tendency will have very big social and economic implications and will lead toward a more effective government. AI applications will not work independently of human beings, at least not in the near future. More likely people will use AI software to help them when analyzing complex problems and taking critical decisions. In most situations, the information generated by AI systems will be complex, and human beings will need assistance to understand it because of the limitations of the human brain. Mind mapping will be the tool of choice for humans working side by side with AI systems. Physicians, engineers, software developers, teachers, law enforcement agencies employees, and many other types of civil servants will enormously benefit from the integration of the mind mapping technique into AI systems. We have to envision a future in which AI systems will enhance our human capabilities, and mind mapping will optimize our use of time.

REFERENCES

Aalto University Documents, January 30, 2017. Available from: http://www.aalto.fi/en/.

Agichtein, E., Gravano, L., 2000. Snowball: extracting relations from large plain-text collections. In: DL '00 Proceedings of the Fifth ACM Conference on Digital Libraries, San Antonio, Texas, pp. 85–94.

Ahuja, J., Webster, J., 2001. Perceived disorientation: an examination of a new measure to assess web design effectiveness. Interacting with Computers 14, 15–29.

Barcelona Carpeta del Ciudadano Documents, January 30, 2017. Available from: http://ajuntament.barcelona.cat/hisenda/es/carpeta-del-ciudadano.

Barcelona Graduate School of Economics Documents, February 14, 2017. Available from: http://www.barcelonagse.eu/.

Barcelona Graduate School of Economics, December 15, 2016. Master's Degree in Data Science. Available from: http://www.barcelonagse.eu/study/masters-programs/data-science.

BCN Analytics Documents, February 14, 2017. Available from: http://bcnanalytics.com/.

BDMA, February 14, 2017. Erasmus Mundus Joint Master Degree Programme in Big Data Management and Analytics. Available from: http://bdma.univ-tours.fr/bdma/.

BDV Documents, February 14, 2017. Available from: http://www.bdva.eu/.

Ben Hamida, S., et al., August 2016. Value proposition design for systems and services by adapting affordance-based design. In: ASME 2016 International Design Engineering Technical Conferences and Computers and Information in Engineering Conference, Vol. 7: 28th International Conference on Design Theory and Methodology, Charlotte, North Carolina, USA, pp. 21–24.

Big Data Spain Documents, February 14, 2017. Available from: https://www.bigdataspain.org/.

BIGDATACoE Documents, February 14, 2017. Available from: http://www.bigdatabcn.com/en/.

Bismart Documents, February 14, 2017. Available from: https://bismart.com/en/home/.

Bower, G.H., Clark, M.C., Lesgold, A.M., Winzenz, D., 1969. Hierarchical retrieval schemes in recall of categorized word lists. Journal of Verbal Learning and Verbal Behavior 8, 323–340.

Brunschwig, C.R., 2006. Visualizing legal information: mind maps and e-government. Electronic Government, an International Journal 3 (4), 386–403.

BSC Documents, January 30, 2017. Available from: https://www.bsc.es/research-and-development/research-areas/big-data.

Burgess-Allen, J., Owen-Smith, V., 2010. Using mind mapping techniques for rapid qualitative data analysis in public participation processes. Health Expectations 13 (4), 406–415. Available from: http://onlinelibrary.wiley.com/doi/10.1111/j.1369-7625.2010.00594.x/epdf.

Burton, R., 2013. Using Mind Mapping Software to Test Website Usability. Available from: http://assurity.co.nz/community/big-thoughts/using-mind-mapping-software-to-test-website-usability/.

ClinicalTrials.gov Documents, February 15, 2017. Available from: http://www.clinicaltrials.gov/.

CMS.gov, 2014. Medicare Provider Utilization and Payment Data: Physician and Other Supplier. Available from: http://www.cms.gov/Research-Statistics-Data-and-Systems/Statistics-Trends-and-Reports/Medicare-Provider-Charge-Data/Physician-and-Other-Supplier.html.

Corbetta, M., Shulman, G.L., 2002. Control of goal-directed and stimulus-driven attention in the brain. Nature Reviews Neuroscience 3, 201–215.

Cowan, N., 2010. The magical mystery four: how is working memory capacity limited, and why? Current Directions in Psychological Science 19 (1), 51–57.

Data.europa.eu Documents, January 30, 2017. Available from: https://data.europa.eu/.

DATALANDSCAPE, January 30, 2017. The European Data Market Monitoring Tool. Available from: http://www.datalandscape.eu/.

Datos.gob.es Documents, January 30, 2017. Available from: http://datos.gob.es/.

DebateGraph Documents, February 17, 2017. Available from: http://debategraph.org/flash/fv.aspx?r=19224.

Department of Finance, 2013. The Government Data Landscape in Australia. The Government of Australia. Available from: http://www.finance.gov.au/blog/2013/10/26/government-data-landscape-australia/.

Department of Health, 2004. Mind-mapping the Impact of Community Factors. The Government of Australia. Available from: http://www.health.gov.au/internet/publications/publishing.nsf/Content/drugtreat-pubs-front8-fa-toc~drugtreat-pubs-front8-fa-secb~drugtreat-pubs-front8-fa-secb-9~drugtreat-pubs-front8-fa-secb-9-3.

EDF Documents, December 15, 2016. Available from: http://www.data-forum.eu/.

Eindhoven University of Technology Documents, January 30, 2017. Available from: https://www.tue.nl/en/.

EIT Documents, January 30, 2017. Available from: http://eit.europa.eu/.

ESADE Documents, February 20, 2017. Available from: http://www.esade.edu/management/eng/programmes/business-analytics.

EU Data Market SMART 2013/0063, December 15, 2016. D6 First Interim Report. IDC. Available from: https://idc-emea.app.box.com/s/k7xv0u3gl6xfvq1rl667xqmw69pzk790.

EUCLID Documents, February 14, 2017. Available from: http://www.euclid-project.eu/.

European Commission, December 15, 2016a. Big Data. Available from: https://ec.europa.eu/digital-single-market/en/big-data.

European Commission, December 15, 2016b. Connectivity for a European Gigabit Society. Available from: https://ec.europa.eu/digital-single-market/en/connectivity-european-gigabit-society.

European Commission, December 15, 2016c. Data Science. Available from: https://ec.europa.eu/digital-single-market/en/news/communication-data-driven-economy.

European Commission, December 15, 2016d. European Legislation on Reuse of Public Sector Information. Available from: https://ec.europa.eu/digital-single-market/en/legislative-measures.

European Commission, December 15, 2016e. Non-legislative Measures to Facilitate Re-use. Available from: https://ec.europa.eu/digital-single-market/en/non-legislative-measures-facilitate-reuse.

European Commission, December 15, 2016f. Open Data. Available from: https://ec.europa.eu/digital-single-market/en/open-data.

European Commission, December 15, 2016g. Open Data Portals. Available from: https://ec.europa.eu/digital-single-market/en/open-data-portals.

European Data Landscape Monitoring Tool Documents, February 10, 2017. Available from: http://www.datalandscape.eu/european-data-market-monitoring-tool.

European Data Portal, 2015. Open Data Maturity in Europe. Available from: http://www.europeandataportal.eu/sites/default/files/edp_landscaping_insight_report_n1_-_final.pdf.

European Data Portal Documents, January 30, 2017. Available from: https://www.europeandataportal.eu/.

European Data Science Academy Documents, December 15, 2016. Available from: http://edsa-project.eu/.

European Institute of Innovation, Technology (EIT) Digital Master School, January 30, 2017. Data Science Master. Available from: https://www.masterschool.eitdigital.eu/programmes/dsc/.

Farrand, P., Hussain, F., Hennessy, E., 2002. The efficacy of the mind map study technique. Medical Education 36 (5), 426–431.

Fillion, F.M., Boyle, C.D.B., 1991. Important issues in hypertext documentation usability. In: Proceedings of the 1991 Ninth Annual International Conference on Systems Documentation ACM, Chicago, IL, pp. 59–66.

Fraunhofer IAIS Documents, January 30, 2017. Available from: https://www.iais.fraunhofer.de/en.html.

Fraunhofer Institute Documents, January 30, 2017. Available from: https://www.fraunhofer.de/en.html.

Frey, C., December 15, 2016. Mind Mapping Software Blog, 2015 Mind Mapping Software Trends Survey. Available from: http://mindmappingsoftwareblog.com/survey-report-mind-mapping-software-trends/.

Geographica Documents, February 14, 2017. Available from: https://geographica.gs/en/.

Govexec.com, February 14, 2017. Trump to Issue First Budget in Mid-march, Promising 'A Lot More with Less'. Available from: http://www.govexec.com/management/2017/02/trump-issue-first-budget-mid-march-promising-lot-more-less/135624/.

GOV.UK Service Management, 2016. Exploratory Testing. Available from: https://www.gov.uk/service-manual/technology/exploratory-testing.

GOV.UK Digital Marketplace, February 14, 2017. Transformation Map Requirements. Available from: https://www.digitalmarketplace.service.gov.uk/digital-outcomes-and-specialists/opportunities/1882.

Guerrero, J.M., 2016. Introducción a la Técnica de Mapas Mentales. EUOC, Barcelona.

Guerrero, J.M., Ramos, P., 2015. Introduction to the Applications of Mind Mapping in Medicine. ImedPub, London.

Hemp, P., 2009. Death by Information Overload. Available from: https://hbr.org/2009/09/death-by-information-overload.

Ho, W.-C., Lin, Y.-C., 2013. The application of mind mapping in maintenance knowledge management and sharing. In: The Thirteenth East Asia-Pacific Conference on Structural Engineering and Construction (EASEC-13), Sapporo, Japan, pp. C-3–C-5.

Holland, B., et al., 2004. An Investigation into the Concept of Mind Mapping and the Use of Mind Mapping Software to Support and Improve Student Academic Performance. Available from: http://wlv.openrepository.com/wlv/handle/2436/3707.

Horizon 2020 Documents, December 15, 2016. Available from: https://ec.europa.eu/programmes/horizon2020/.

IBM Alchemy Documents, January 15, 2017. Available from: http://www.ibm.com/watson/alchemy-api.html.

IBM Watson Analytics Documents, January 15, 2017. Available from: https://www.ibm.com/analytics/watson-analytics/us-en/.

IBM Watson Documents, January 15, 2017. Available from: https://www.ibm.com/watson/.

ideXlab Documents, January 30, 2017. Available from: http://www.idexlab.com/en/.

Infoseg, 2006. Dynamic Mind Mapping. Available from: http://www.infoseg.com/gmu/dmm.pdf.

Infoseg, 2013a. Statistical Analysis with R – Slideshare Presentation. Available from: http://www.slideshare.net/jmgf2009/r-and-mind-mapping-automation.

Infoseg, 2013b. Statistical Analysis with R – PNG Sample. Available from: http://www.infoseg.com/gmu/SAMM.png.

Infoseg, 2013c. Open Data – Slideshare Presentation. Available from: http://www.slideshare.net/jmgf2009/open-datamm.

Infoseg, 2013d. Open Data – PNG Sample. Available from: http://www.infoseg.com/gmu/EUDP.png.

Infoseg, 2014a. Big Open Data Analysis Using R – Slideshare Presentation. Available from: http://www.slideshare.net/jmgf2009/big-open-data-in-medicine-with-r-and-mind-mapping.

Infoseg, 2014b. Big Open Data Analysis Using R – PNG Sample. Available from: http://www.infoseg.com/gmu/Medicare.png.

Infoseg, 2014c. Clinical Trials Open Data – Slideshare Presentation. Available from: http://www.slideshare.net/jmgf2009/open-data-in-medicine-application-of-mind-maping-automation-to-visualize-information.

Infoseg, 2014d. Clinical Trials Open Data – PNG Sample. Available from: http://www.infoseg.com/gmu/CT.png.

Infoseg, 2014e. Mind Mapping Automation in the Visualization of SAP Information, – Slideshare Presentation. Available from: https://www.slideshare.net/jmgf2009/mind-mapping-automation-in-the-visualization-of-sap-information.

Infoseg, 2014f. Applications of Mind Mapping Automation in the Analysis of Information Security Log Files, – Slideshare Presentation. Available from: https://www.slideshare.net/jmgf2009/applications-of-mind-mapping-automation-in-the-analysis-of-information-security-log-files.

Infoseg, 2016. Automating the Conversion of Microsoft Dynamics CRM Information into Mind Maps – Slideshare Presentation. Available from: https://www.slideshare.net/jmgf2009/automating-the-conversion-of-microsoft-dynamics-crm-information-into-mind-maps.

Infoseg, January 3, 2017. Visualization of IBM Watson Complex Information as Mind Maps – Slideshare Presentation. Available from: http://www.slideshare.net/jmgf2009/mind-mapping-to-visualize-complex-information-generated-by-ibm-watson.

Infoseg Documents, March 3, 2017. Available from: http://www.infoseg.com/mi_01_en.shtml.

Innaxis Documents, February 14, 2017. Available from: http://www.innaxis.org/.

Insight Public Affairs (IPA), 2012. Infrastructure Policy Mind Map. Available from: http://insightpublicaffairs.com/2012/10/infrastructure-policy-mind-map/.

Instituto de Empresa Documents, February 14, 2017. Available from: http://www.ie.edu/.

Instituto de Empresa, December 15, 2016. Master in Business Analytics and Big Data. Available from: http://bigdata.ie.edu/.

Jožef Stefan Institute Documents, February 14, 2017. Available from: http://www.ijs.si/ijsw/JSI.

Kungliga Tekniska Högskolan Documents, February 14, 2017. Available from: http://www.kth.se/en.

Larkin, J.H., Simon, H.A., 1987. Why a diagram is (sometimes) worth ten thousand words. Cognitive Science 11, 65–99.

Lyon, D.C., 2007. The evolution of visual cortex and visual systems. In: Kaas, J.H. (Ed.). Kaas, J.H. (Ed.), Evolution of Nervous Systems, vol. 3. Elsevier, London, pp. 267–306.

MacDonald, J., 2015. 6 Steps to Strategy Using Mind Mapping. Available from: http://www.business2community.com/strategy/6-steps-strategy-using-mind-mapping-01124576#tmvqRQ8PTcZkgjz5.97.

McConnachie, D., 2009. Mind Mapping Keeps Defence Projects on Track. Available from: http://www.computerworld.com.au/article/314512/mind_mapping_keeps_defence_projects_track/.

Microsoft CRM Documents, March 12, 2017. Available from: https://www.microsoft.com/en-us/dynamics365/what-is-crm.

Microsoft Sharepoint Documents, March 12, 2017. Available from: https://products.office.com/en-us/sharepoint/collaboration.

Miller, G.A., 1956. The magical number seven, plus or minus two: some limits on our capacity for processing. Psychological Review 63 (2), 81–97.

Mind Mapping at Boeing: Getting Off the Ground, December 23, 2016. (video file). Available from: https://www.youtube.com/watch?v=qHIPqBfKa0g.

Mindjet Documents, March 12, 2017. Customers – Fortune 100. Available from: https://www.mindjet.com/ja/customers/.

MindManager Documents, March 12, 2017. Available from: https://www.mindjet.com/mindmanager-windows/.

MindManager Enterprise Documents, March 12, 2017. Available from: https://www.mindjet.com/uses/mindmanager-enterprise/.

Mongin, P., 2005. Comprendre la situation dans des quartiers difficiles. Available from: https://www.petillant.com/Comprendre-la-situation-dans-des.

Nelson, D.L., Reed, U.S., Walling, J.R., 1976. Pictorial superiority effect. Journal of Experimental Psychology: Human Learning and Memory 2, 523–528.

Nguyen, V., Nguyen-Viet, H., Pham-Duc, P., Wiese, M., 2014. Scenario Planning for Community Development in Vietnam: A New Tool for Integrated Health Approaches? Available from: https://www.ncbi.nlm.nih.gov/pmc/articles/PMC4138499/.

Open Data gencat Documents, January 30, 2017. Available from: http://dadesobertes.gencat.cat/en/index.html.

Open Data Incubator for Europe (ODINE), January 30, 2017. Available from: https://opendataincubator.eu/.

Open Data Institute Documents, February 14, 2017. Available from: http://theodi.org/.

Open Knowledge Foundation Documents, January 30, 2017. Available from: https://okfn.org/.
OpenDataBCN Documents, January 30, 2017. Available from: http://opendata.bcn.cat/opendata/en.
Persontyle Documents, January 30, 2017. Available from: https://www.persontyle.com/.
Piperlab Documents, January 30, 2017. Available from: http://piperlab.es/en/.
Politecnico di Milano Documents, January 30, 2017. Available from: http://www.polimi.it/en/university/.
Polytech'Nantes Documents, January 30, 2017. Available from: http://web.polytech.univ-nantes.fr/.
SAP Documents, March 12, 2017. Available from: https://www.sap.com/index.html.
Stone, L., 2006. Continuous Partial Attention. Available from: https://lindastone.net/qa/continuous-partial-attention/.
Synergic Partners Documents, February 14, 2017. Available from: http://www.synergicpartners.com/en/.
Technische Universität Berlin Documents, January 30, 2017. Available from: http://www.dima.tu-berlin.de/menue/teaching/international_programs/eit_digital_data_science_master.
Telefonica, December 15, 2016. Master Telefonica in Big Data. Available from: http://www.campusbigdata.com/master-telefonica-en-big-data-y-business-analytics/.
Telefonica Documents, February 14, 2017. Available from: https://www.telefonica.com/en/.
Telefonica Open Future_Documents, January 30, 2017. Available from: https://www.openfuture.org/en.
The EU Open Data Portal Documents, January 30, 2017. Available from: http://data.europa.eu/euodp/.
Toffler, A., 1970. Future Shock. Random House, New York.
United States Incident Reporting System (US-CERT), December 18, 2016. Available from: https://www.us-cert.gov/forms/report.
Universidad Juan Carlos I, December 15, 2016. Master in Data Science. Available from: http://www.masterdatascience.es/.
Universidad Juan Carlos I Documents, February 14, 2017. Available from: http://www.urjc.es/en/.
Universidad Politécnica de Madrid Documents, January 30, 2017. Available from: http://www.upm.es/internacional.
Università degli Studi del Piemonte Orientale Documents, January 30, 2017. Available from: http://www.unipmn.it/default.aspx?r2b_language=Eng&oid=&oalias=.
Universitat de Barcelona de Barcelona Documents, January 30, 2017. Available from: http://studies.uoc.edu/en/study-at-the-uoc.
Universitat Oberta de Catalunya Documents, January 30, 2017. Available from: http://studies.uoc.edu/en/study-at-the-uoc.
Universitat Oberta de Catalunya, December 15, 2016. Master in Business Intelligence and Big Data. Available from: http://estudios.uoc.edu/es/masters-posgrados-especializaciones/master/informatica-multimedia-telecomunicacion/inteligencia-negocio-big-data/presentacion.
Universitat Politècnica de Catalunya Documents, January 30, 2017. Available from: http://www.upc.edu/?set_language=en.
Université de Lyon Lumière Lyon 2 Documents, January 30, 2017. Available from: http://www.univ-lyon2.fr/www0-home-157320.kjsp?RF=WWW_EN.
Université François Rabelais Tours Documents, January 30, 2017. Available from: http://international.univ-tours.fr/welcome-international-265902.kjsp?RH=INTER&RF=INTER-EN.
Université Libre de Bruxelles Documents, January 30, 2017. Available from: http://www.ulb.ac.be/ulb/presentation/uk.html.
Université Nice Sophia Antipolis Documents, January 30, 2017. Available from: http://unice.fr/en.
Université Pierre and Marie Curie Paris 6 Documents, January 30, 2017. Available from: http://www.upmc.fr/en/.

US Census Bureau, 2015. Available from: https://factfinder.census.gov/faces/tableservices/jsf/pages/productview.xhtml?src=bkmk.

Van Oudsheuden, M., 2013a. Mind Mapping in Large Government Departments – Video. Available from: https://www.youtube.com/watch?v=9FdfI8TekoQ.

Van Oudsheuden, M., 2013b. Mind Mapping in Large Government Departments – Mind Map. Available from: http://www.slideshare.net/connectionofminds/03-mind-mapping-within-large-government.

Ware, C., 2004. Information Visualization. Perception for Design. Elsevier-Morgan Kaufman, San Francisco.

Wu, Z.-Y., Wang, X.-A., Wu, Y., 2009. E-government system demand analysis based on mind map. In: Proceedings of the 2009 International Conference on Networking and Digital Society, Guiyang, China, pp. 254–257.

Section 3

Federal Data Science Use Cases at the US Government

Chapter 9

A Deployment Life Cycle Model for Agricultural Data Systems Using Kansei Engineering and Association Rules

Feras A. Batarseh, Ruixin Yang
George Mason University, Fairfax, VA, United States

If a machine is expected to be infallible, it cannot also be intelligent.

Alan Turing

1. INTRODUCTION AND BACKGROUND

One of the most prominent questions in any software development project: *Is the system ready for deployment?* A very simple question that has a very complicated answer. Answering the question is not only important to the development team but also crucial for project planning, budgeting, resource allocation, and many other aspects that affect the organization and federal employees involved in the project. More precisely, when software development takes place at a federal agency, there are many constraints and challenges that are specific to federal software development (such as: politics, bureaucracy, the view of the public, the media, and monetary issues–to name a few). Having errors in a system certainly does not help; however, errors in *intelligent* systems do not directly dispose them as *unintelligent* (Turing's inspirational quote at the beginning of this chapter). Many argue that errors are inevitable, but how are they caught, minimized, and fixed? Furthermore, how is user acceptance insured? Many methods have been introduced in literature; however, it is a special case when dealing with federal systems. The answers to all those issues are presented in this chapter.

Federal Data Science. http://dx.doi.org/10.1016/B978-0-12-812443-7.00009-0

1.1 A Measuring Stick

The obvious scientific approach to answering questions in software projects is through *measuring* and quantifying. In software development, that "quantitative" approach for capturing errors is called validation and verification (V&V). Similar to other software life cycle development phases, testing (which consists of V&V) is a science and an art (Batarseh, 2012). Testing is considered one of the most important phases in most (if not all) development life cycles.

Previously, software development at the government had followed a waterfall model (Sommerville, 1982), where everything falls downward that incurred massive costs and time delays. Based on an article by the Washington Post (DePillis, 2013), monetary values of up to $50 million per project are wasted on certain failing federal projects. Unfortunately, many governmental agencies still latch onto the expensive and difficult to manage waterfall model (such as: the Financial Industry Regulatory Authority and the Department of Health and Human Services). Overall, federal agencies spend around $76 billion dollars on software and IT every year (Pace Systems, 2017).

One of the recent and known "waterfall" software failures was the Affordable Care Act website (www.healthcare.gov) (Smith and Cohen, 2013). A failing federal system or government website for programs that are applied across the country could have a mammoth effect on the direction the country takes in certain policies; it could also exacerbate public frustration, draw major negative media focus onto the issue, and put the administration under a lot of scrutiny or criticism. Therefore, testing federal systems is an unavoidable and critical process that needs to be performed diligently.

1.2 Motivation

The United States Government Accountability Office (GAO) published a report that spotlighted on the issue of software failures at government (GAO Reports, 2012). The GAO report identified 14 issues that governmental agencies need to solve. To address the listed 14 challenges, GAO defined 32 practices that the government needs to adopt for better deployments. Most of these federal issues are *life cycle–related* problems. This is the main motivation of this chapter: **to illustrate that a well-defined life cycle model is able to mitigate many of the common federal software development processes, and lead to better software, and eventually better policy making** (Yu Ming-Tun, 2012).

This chapter (and in fact this book) presents solutions and guidelines that are specific for federal data systems specifically. This chapter presents a new federal life cycle. Therefore, the focus of the next section is a review of existing *life cycles* for data analytics. To test the proposed life cycle, it was deployed at agencies that are part of the United States Department of Agriculture. It focuses on three main aspects: **testing**, **deployment**, and **user satisfaction**–all are discussed next.

1.3 Systems Life Cycle Models

Famous life cycle models include: waterfall model, spiral model, incremental model, agile model, user-oriented model, and many others. Life cycles in software engineering have led major shifts in the tech progress of the engineering world. The transition from waterfall development to spiral, from traditional to agile, and from product-based to contextual user-based development left an obvious fingerprint on the world of engineering (Batarseh and Gonzalez, 2011; Dalal and Chhillar, 2012). Software engineering, like all other engineering disciplines, follows a set of processes to construct a system. The software processes directly influence how the system is incepted. These processes (when put together) result in a life cycle. Multiple software engineering life cycle models circulate in literature, but it is evident that the death of the traditional system life cycle model may be very close or has already happened. Novel life cycle models (enabled by the web and distant collaborations within open source development communities) are continuously shifting business processes in response to these conditions and giving rise to a new generation of software life cycles. These new models are very different in that they are **incremental**, **agile**, **iterative**, **spiral**, **intelligent**, and **reactive** to social and business circumstances (McCall et al., 1977). Most software life cycle models include a variation of the following steps because this is not a contribution of this chapter, the steps are quoted verbatim from (Batarseh and Gonzalez, 2011): "System planning: feasible systems replace or supplement existing information processing mechanisms whether they were previously automated, manual, or informal. Requirement Analysis and Specification: identifies the problems a new software system is supposed to solve, its operational capabilities, its desired performance characteristics, and the resource infrastructure needed to support system operation and maintenance. Functional Specification or Prototyping: identifies and potentially formalizes the objects of computation, their attributes and relationships. Partition and Selection: given requirements and functional specifications, divide the system into manageable pieces that denote logical subsystems. Architectural Design and Configuration Specification: defines the interconnection and resource interfaces between system subsystems. Detailed Component Design Specification: the procedural methods through which the data resources within the modules of a component are transformed from required inputs into outputs. Component Implementation and Debugging: codifies the preceding specifications into operational source code implementations and validates their basic operation. Software Integration and Testing: affirms and sustains the overall integrity of the software system architectural configuration through verifying the consistency and completeness of implemented modules. Documentation Revision and System Delivery: packaging and rationalizing recorded system development descriptions into systematic documents and user guides, all in a form suitable for dissemination and system support. *Deployment and Installation*: providing directions for installing the delivered software into the local computing environment. Training and Use: providing system users with instructional aid and guidance for understanding the system's capabilities and limits in order to effectively use the system. Software Maintenance: sustaining the useful operation of a system in its host/target environment by providing requested functional enhancements." Deployment, installation, and user acceptance are usually not given that much attention, that is the focus of the life cycle presented in this chapter.

1.4 Analytical Models for the Government

The challenge of intelligent and accurate predictions in systems that was previously pursued by many (Koh and Gerald, 2001; Jing, 2004; Groth, 1999) is now being transformed into a new testament of big data and data mining. The goal of data mining is to return an intelligent and more focused version of large and impersonal data sets. Literature has many examples of successful applications of data mining, not only to specific business-driven functions but also many industrial and research domains such as healthcare (Koh and Gerald, 2001), education (Jing, 2004), banking and finance (Groth, 1999), and many others. Undoubtedly, data science is a buzz word. Although considered a fairly novel field, there are some prominent life cycle models for data analytical systems. In this chapter, we make the case for three main phases in a life cycle for federal use: (1) An inevitable phase that traditional life cycles and data engineering life cycles have in common is the need for *testing* (as its already established in previous sections). (2) Federal regulations put *data validity* at the forefront of its priorities; therefore, a life cycle model without comprehensive validation is not sufficient. (3) Successful *deployment* and users' *adoption* are very important aspect as well. If the federal employees do not adopt the tools and models that are presented here, then again, the efforts put in these systems go in vain.

Based on Fig. 9.1 (Data Analytics Handbook, 2006), data analytics tools have a life cycle that starts with all kinds of data cleaning, architecting, and

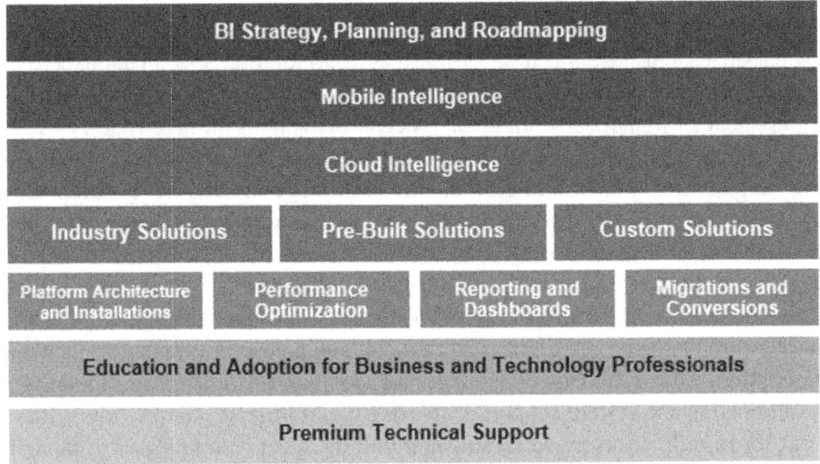

FIGURE 9.1 Data analytics tools (Data Analytics Handbook, 2006).

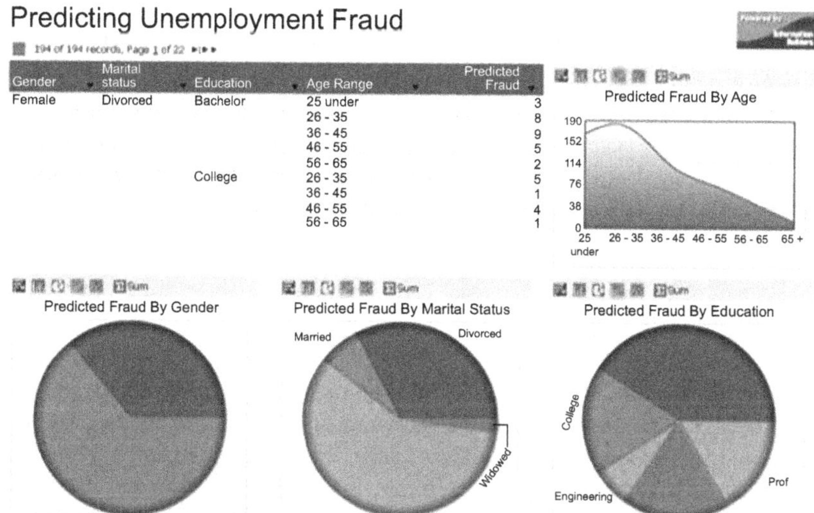

FIGURE 9.2 Example dashboard from a data analytics tool (Data Analytics Handbook, 2006).

modeling, however, in most cases the end user is exposed to data visualizations (reports and dashboards). As part of the analytical process, after the data's architecture is in place and the data are modeled and well structured, a commonplace outcome from a data system is an interactive dashboard that allows for regular users to interact with advanced analytics without understanding the underlying complex algorithms or database SQL queries or any technical detail of that sort. An example of a dashboard that is built for federal users is presented in Fig. 9.2. In the example shown, the dashboard provides decision-making support through a prediction of unemployment fraud rates by different gender sectors (gender, marital status, and education). These dashboards are usually driven by customer requirements and are highly affected by users' acceptance. In the scope of this chapter, Kansei engineering (KE) methods are used to evaluate the deployment efforts and the users' adoption rates of data analytics tools interfaces.

There is a gap in literature for *in-house* federal-driven models to guide the testing and acceptance of a data system. To be successful, such models need to accommodate the following very important five characteristics:

1. Include testing, deployment, and user adoption methods.
2. Are not commercial or vendor-specific.
3. Could be implemented at any federal agency (keeps the user in mind).
4. Are controlled by the agency itself—not dependent on external sources.
5. Support a well-defined deployment process for guiding the federal engineers from A to Z.

Based on what have been discussed in previous sections, and because of the immediate need for a life cycle that has the aforementioned characteristics, this chapter introduces: **The Federal Deployment and Adoption Life cycle (FEDAL)** to address all the challenges and fill the existing void. FEDAL consists of a module for testing that uses *association rules* (AR) to locate errors and fix them (method is called: Association rules testing (**ART**)). FEDAL also consists of a user acceptance module that uses KE to evaluate deployment of a system (method is called: Kansei engineering deployment and traceability (**KDT**)). The two methods within FEDAL (each has a dedicated section below) are presented.

2. RELATED WORK

Turing tests were proposed by Alan Turing (1950) as a method to determine the intelligence of a computing machine when compared to a human being; where a machine and a human played a game anonymously. The tests apply the concepts of anonymity to the validation process. Since then, intelligent software testing has been undeniably one of the most important phases in software development.

2.1 Intelligent Software Testing

Many testing methods have been applying predictive V&V (a form of testing that compares previous validation results with corresponding results). When predictive validation is used, the results are saved, and for the next iteration the same sets of tests are executed and compared with previous results. More methods also applied artificial intelligence (AI) concepts to testing. For example, Afzal et al. (2009) evaluated five different techniques for predicting the number of faults in testing, using the following: (1) particle swarm optimization-based artificial neural networks, (2) artificial immune recognition systems, (3) gene expression programming, (4) genetic programming, and (5) multiple regression. Moreover, one of the newest comprehensive methods that deployed analytics into software testing is analytics-driven testing (ADT), shown in Fig. 9.3 (Batarseh and Gonzalez, 2015).

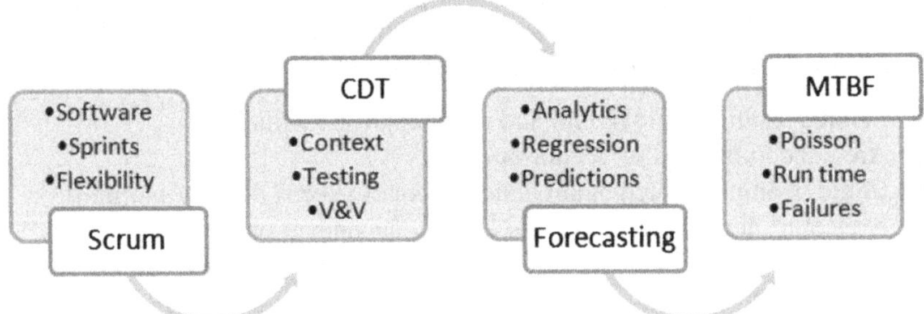

FIGURE 9.3 The analytics-driven testing (ADT) pillars (Batarseh and Gonzalez, 2015). *CDT*, context-driven testing; *MTBF*, mean time between failures.

The main pillars of ADT are: (1) Scrum, (2) context-driven testing, (3) regression forecasting, and (4) mean time between failures (MTBF). Each of these four pillars includes three major drivers of ADT. MTBF regression forecasting is used to forecast failures through incremental Scrum development (Batarseh and Gonzalez, 2015). Other famous modern methods include: (1) Statistical testing using automated search: searching for defects in software systems using summary statistics. (2) Neural networks applied to software cost estimation: measuring cost of software development using AI. (3) Machine learning (ML) for defect prediction: predicting software failures using ML. One can easily think of a prediction system as a *probabilistic reasoner*. One can also think of Bayesian models as learners and of classifiers as learners, probabilistic reasoners, and/or optimizers. Indeed, all of the ways in which AI has been applied to software engineering can be regarded as ways to optimize either the engineering process or its products and, as such, they are all examples of *search-based software engineering*. Some methods applied data mining to software engineering, but none used AR to locate defects in an intelligent system suitable for the federal government.

Historically, the lack of software testing has caused many critical and non-critical projects to fail, human injuries to occur, financial loss to be endured, among many other negative outcomes. One of the known examples of software failures is: "What was to be the world's largest automated airport baggage handling system, became a classic story in how technology projects can go wrong. Faced with the need for greater airport capacity, the city of Denver elected to construct a new state of the art airport that would cement Denver's position as an air transportation hub. Despite the good intentions the plan rapidly dissolved as underestimation of the project's complexity resulted in snowballing problems and public humiliation for everyone involved. Thanks mainly to problems with the baggage system, the airport's opening was delayed by a full 16 months" (Dalal and Chhillar, 2012). Another example is: "A few minutes after the Shuttle Endeavour reached orbit for STS-126 on November 14, 2008, mission control noticed that the shuttle did not automatically transfer two communications processes from launch to orbit configuration. Primary communications continued to use S-band frequencies after they should have transferred to the more powerful Ku-band. The link between the shuttle and its payload—the Payload Signal Processor (PSP)—remained configured for a radio link rather than switching automatically to the hardwired umbilical connection" (McCall et al., 1977).

More examples were provided by the National Aeronautics and Space Association (www.nasa.gov), in 1993, three people died because of a software failure on Airbus A320. More drastically, in 1996, 160 people died in the Ariane 5 Flight 965 (Table 9.1).

Software testing is the most common approach to eliminate software defects and eventually eliminate disasters like the ones mentioned. Testing is conceptually divided into two main parts: V&V. Validation is the process of ensuring that the developed system addresses the presented software requirements, whereas

TABLE 9.1 Famous Examples on Software Failures Losses (Dalal and Chhillar, 2012)

	Airbus A320 (1993)	Ariane 5 Galileo Poseidon Flight 965 (1996)	Lewis Pathfinder USAF Step (1997)	Zenit 2 Delta 3 Near (1998)	DS-1 Orion 3 Galileo Titan 4B (1999)
Aggregate cost		$640 million	$116.8 million	$255 million	$1.6 billion
Loss of life	3	160			
Loss of data		Yes	Yes	Yes	Yes

TABLE 9.2 The Continuous Shift in Focusing on Testing (McCall et al., 1977)

	Requirements Analysis	Preliminary Design	Detailed Design	Coding and Unit Testing	Integration and Test	System Test
1960s–70s	10%			80%	10%	
1980s	20%		60%		20%	
1990s	40%	30%		30%		

verification aims to eradicate defects from the system (such as syntax and runtime errors). Throughout the years, V&V have been performed in many different ways; however, one thing that most scientists agree on is the increased need to perform "more" testing. Thereafter, there has been serious and continuous shifts in the percentage of time/effort spent on the testing phase of the development life cycle. The National Institute of Standards and Technology (NIST) published a comprehensive study on the recent trends of software testing. Software development and coding resource consumption have been reduced (from 80% in the 1960s to 30% in the 1990s), and therefore, more time has been assigned to testing (from 10% to 30%) and other phases (Table 9.2). Testing constantly pursues both a valid and a verified software system. Testing aims to check as many "ility" boxes as possible; Reliability, Usability, Flexibility, and Reusability, are among the most desired goals of testing. McCall and other scientists introduced a list of quality factors for software, shown in Table 9.3.

TABLE 9.3 Overall Software Quality Factors (NIST, 2015)

Attribute	Description
Product Operation	
Correctness	How well the software performs its required function and meets customers' needs
Reliability	How well the software can be expected to perform its function with required precision
Integrity	How well accidental and intentional attacks on the software can be withstood
Usability	How easy it is to learn, operate, prepare input of, and interpret output of the software
Efficiency	Amount of computing resources required by the software to perform its function
Product Revision	
Maintainability	How easy it is to locate and fix an error in the software
Flexibility	How easy it is to change the software
Testability	How easy it is to tell if the software performs its intended function
Product Transition	
Interoperability	How easy it is to integrate one system into another
Reusability	How easy it is to use the software or its parts in other applications
Portability	How easy it is to move the software from one platform to another

The many testing challenges lead to the need of standardizing software quality factors, and therefore, an ISO standard for testing was introduced (ISO 9126) that includes attributes such as portability, reliability, functionality, usability, efficiency, and maintainability (ISO Standards 9126, 2016). To address these factors, testing has been done in many ways, shapes, and forms. However, testing processes are categorized into three major parts: (1) General testing, such as: subroutine/unit testing, regression tests, integration testing, and system testing. (2) Specialized testing, such as stress testing, recovery testing, security testing, and survivability testing. (3) User-involved testing, such as usability testing, field testing, acceptance testing, and alpha/beta tests. Based on the recent study by NIST (2015), the ability to define defect locations is listed as the *most time-consuming* activity of software testing. In the study, the NIST researchers compiled a vast number of software projects and reached the following conclusion:

"If the location of bugs can be made more precise, both the calendar time and resource requirements of testing can be reduced. Modern software products typically contain millions of lines of code. Precisely locating the source of bugs in that code can be very resource consuming. Most bugs are introduced at the unit stage. Thus, effective testing methods for finding such bugs before units are combined into components and components into systems would be especially valuable." Literature has a gap in defect traceability; no AI-driven method was introduced to help analysts and engineers pinpoint defects that would affect the overall health of software. In this chapter, this gap is filled by applying AR (through a method called ART) to trace defects intelligently within federal data systems.

2.2 Kansei and Software Deployment (A Review)

Traditionally, the success of a software design depends on the designers' artistic sensibilities, which quite often did not exactly meet what the user desires. Therefore, many product design studies have been carried out to get a better insight into consumers' subjective perceptions. The most notable research to address such issue is KE (Nagamachi, 1995). KE principles were first introduced in Japan in the 1970s. Most of the product design research community was exposed to KE after the publication of a famous journal chapter by Nagamachi (1995). The idea of KE is based on the notion that a physical form of a product plays a major role in formulating customers' buying decisions that led multiple researchers to study what forms or features may or may not affect customers' responses, accordingly design software systems to appeal to the users and guarantee a positive user response. For example, with the recent rise in popularity of online shopping, the visual appeal of a product is essential in today's software's marketplace. Users interact with a wide array of products during the course of their lives and subconsciously develop solid product evaluation and emotional discrimination skills. Based on that, the task of software designers and testers is to manipulate the product form features (PFFs) to embed specific styles which satisfy the consumer's expectations into the software. The goal is to satisfy the consumers' affective responses (CARs) in a positive manner to increase software acceptance. The relation between CARs, PFFs, and the users is shown below in Fig. 9.4. There is a general lack of research in the field of

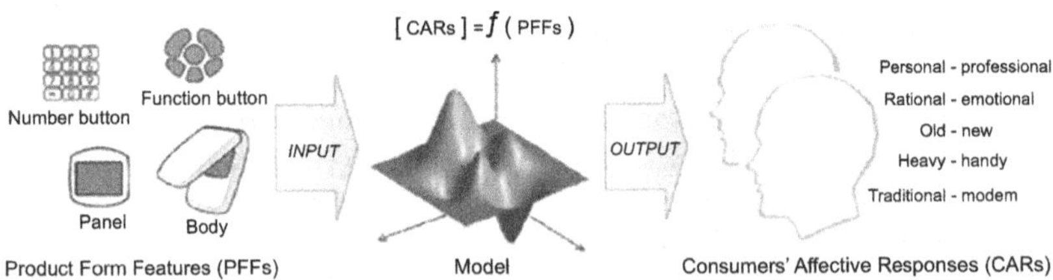

FIGURE 9.4 Kansei engineering (Nagamachi, 1995).

KE. However, one famous model that is based on KE is the Kano model. The model was created in 1984 by Noriaki Kano while studying the contributing factors to customer satisfaction and customer loyalty (Yang, 2011; Hartono and Chua, 2011). Kano classified five unique categories of customer requirements, three of which developers would want to have in the offering, the other two should be taken out (in most cases). This model's main objective is to help teams understand, classify, and integrate three main categories of requirements into the products that they are developing.

The five categories of customer requirements are classified depending on their ability to create customer satisfaction or cause dissatisfaction. Knowing what category customer requirements fall into and the importance of each requirement can help prioritize development activities and determine what to include in the offering and where to spend resources improving these requirements. In 2012, Shi and Sun deployed Kansei to mining through rule bases, critical form features, and KE parameters were defined as condition attributes and decision attributes, respectively, which were formalized in what they called in their chapter as decision tables. They also presented the semantic differential, which measures the connotative meaning of KE concepts and is applied to evaluate form features of the product through a KE questionnaire system (Shi et al., 2012).

None of the work that was found in literature was designed for software, except for the work of Almagro and Martorell in 2012; they published a KE quality assurance model that is based on comparing the semantic space of Kansei and the space of software properties (Almagro and Martorell, 2012). They use a synthesizing engine to achieve an understanding of customer's feedback. Other published general methods to capture a Kansei include: semantic descriptions of environments (textual descriptions of user's sentiments), quality function deployment (a generic quality control function that gets applied during product development), and conjoint analysis (analysis that is done by multiple parties to cover different aspect of a product). It is clear then, that there is a gap in applying KE concepts to software engineering, in this chapter, Kansei is applied through KDT, in a method aiming to fill the void in literature. In KDT, CARs are driven by the federal analyst's continuous feedback and the analysis of their sentiments.

3. THE FEDERAL DEPLOYMENT AND ADOPTION LIFE CYCLE

As previously presented, FEDAL is a deployment life cycle that includes **two** main parts: **ART** and **KDT**. Both parts are introduced next.

3.1 Association Rules Testing

ART is based on AR. A commonplace statistical model that aims to establish relations (associations) between different data points. In the case of federal systems testing, ART associates errors that occur in a federal system, to help the engineer predict where the next error will occur. At each iteration of training of the AR model, data are retrained, and the model yields different outcomes. In ART, the engineer cares about the resulted predictions (called consequents in AR); he/she needs to use them to locate defects in next iterations of development. ART consists of five spiral steps and each step is performed at a different stage of the spiral workflow. Step 1 is always performed at the beginning of the iteration. Step 2 is considered based on the amount of data collected. The engineer asks the following question: will the new collected data affect the outcome of the model? If sufficient amount of data records is collected that requires retraining of the model, then Step 2 is executed. Otherwise, Step 3 is carried out. Step 4 aims to split the "good results from the bad results." Using confidence (which measure the "goodness" of the model), the results are sorted, and the engineer aims to find the highest confidence, with high support. Step 5 consists of using the highest sorted consequences (which represent system modules) for tracing errors in the software system. The five steps are summarized below:

1. ART defect data collection: To increase the accuracy of any data mining model, a fair amount of high quality data is required.
2. ART model development and training: In this stage, the AR model is built (using an analytical tool such as Tableau, SAS, or SPSS), and the data are trained.
3. AR model results: Outputs of the model are antecedents, consequents, confidence, and support (these are discussed in detail next).
4. Sort all ART predictions: Consequent predictions are sorted by confidence, top predictions with highest confidence results are then considered for error tracing.
5. ART intelligent testing: using the outcomes from Step 3 and the predictions from Step 4, testing is performed and focused on the system modules presented in the AR model's outcomes. Then go back to Step 2 (training).

All five stages are presented in Fig. 9.5. Step 1 is blue, 2 is orange, 3 is green, 4 is purple, and 5 is red. Note that Step 1 is always executed. Steps 2, 3, and 4 are not; depending on how Step 5 is performed, and only after two or three spiral iterations occur. Before the model is trained with sufficient data, it cannot be trusted. After Step 5, the engineer should be able to identify the following: (1) the riskiest module(s) in the federal system, (2) the most common location of defects, and (3) what defects will occur next and where. The experimental work section presents a complete example on the process, with results and discussions. Please refer to web version for a colored figure.

3.2 Kansei Engineering Deployment and Traceability

KDT is the second part of FEDAL, besides ART. KDT is applied to get the federal analysts feedback (especially nontechnical staff, such as agricultural

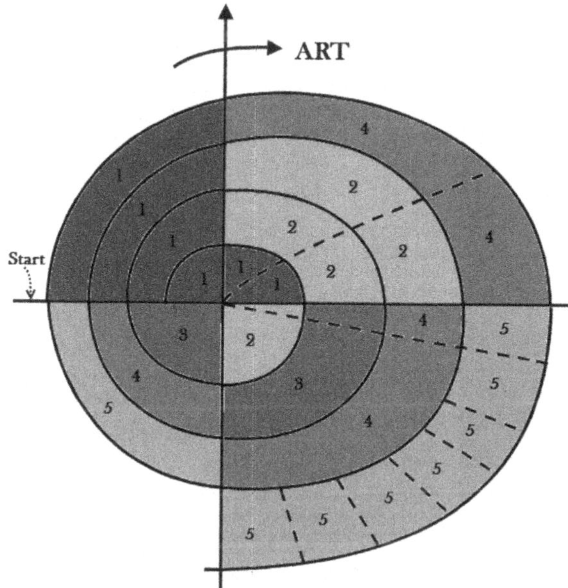

FIGURE 9.5 The proposed spiral association rules testing (ART) method.

analysts). That is done either after the system is deployed or during development. The main parts of the KDT method are:

- A rule base: when users of a certain federal agency start using a data system, they get access to the KDT tool to insert feedback about the software into the rule base (using SQL insert statements in the background). The tool gathers survey inputs from the users and uses the sentiment analysis (SA) code to measure sentiments regarding the software (all CARs are saved in this part of KDT).
- SA algorithm: the SA code is written in the statistical language, R, using the following libraries: library (*tm*) and library (*RTextTools*). Sentiment is measured through a list of predefined words with weights that range between −5 and 5. A word that is very negative gets a score of −5, words that reflect no sentiments are given a score of 0, whereas score 5 reflects a very positive sentiment. The matching engine then uses these sentiments to connect them to the CAR rules in the rule base and guides the engineers through the required changes. An initial default score of 0 is associated with each rule. These scores are updated regularly as federal users are interacting with the tools and the dashboards.
- The Kansei matching engine: The matching engine looks for sentiment scores and associates them with rules in the rule base (all PFFs of KE are saved in this engine). The sentiment scores lead to changes in the rule base. The KDT process is based on four main pillars: Kansei matching engine, the rule base,

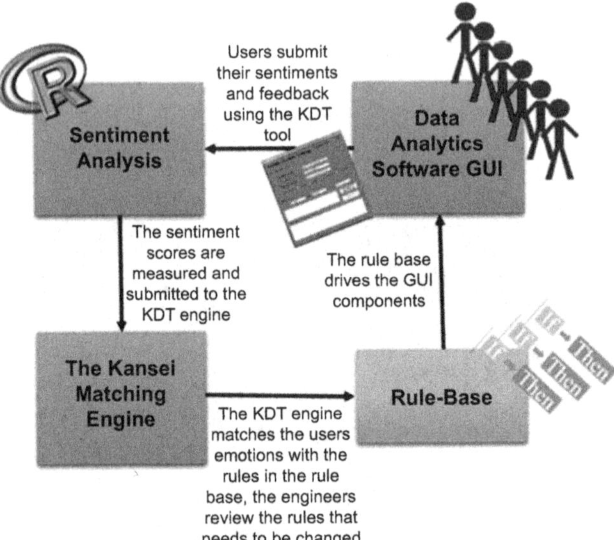

FIGURE 9.6 The Kansei engineering deployment and traceability (KDT) process for agricultural analysts.

SA using R, and the federal data analytics tools. The five steps of KDT are (also refer to Fig. 9.6 for an illustration):

1. Create a rule base for the federal system in SQL. Rules have the following simple format:
   If (----) then (----) else (----) then a user starts using the system.
2. Users log issues and notes for their sentiments of the tool.
3. The SA R code provides sentiment scores for users' inputs.
4. The matching engine associates scores with the rules in the SQL rule base and changes the rules scores based on users' sentiments. A number of words are associated with each rule, and the engine searches for those words in the user's text to match rules with sentiments. The sentiment score of each rule is continuously updated based on users' inputs.
5. Rules with low scores are updated and possibly recreated (by the engineers). When another user starts using the tool or the dashboards, the KDT process goes back to Step 2.

4. EXPERIMENTAL STUDIES ON KANSEI ENGINEERING DEPLOYMENT AND TRACEABILITY AND ASSOCIATION RULES TESTING

This section introduces the experimental work used to evaluate FEDAL (KDT and ART) using federal agricultural data. Results are presented and discussed next.

4.1 Code Coverage and Maintenance Costs of Association Rules Testing

An important aspect at most federal agencies is the cost of a project (both development and maintenance costs). Through ART, to reduce the running costs of a federal system, it is important to deliver a system that is fully validated. One of the main validation assessment metrics is code coverage. In one of our federal projects, Microsoft's Visual Studio (which has a code coverage functionality–see Fig. 9.7) is used to measure the coverage percentage for a system tested with ART, and another one tested with a traditional waterfall model.

The coverage percentages and the associated maintenance costs were collected at the federal agency. As it is shown in the previous results, big data validation was a priority through ART. In ART, a total of 11 months were assigned to testing, whereas with waterfall models, only 2 months of after-the-fact testing took place. When a waterfall model is implemented, it becomes very difficult to inject incremental testing into it, and the engineers are forced to run tests at the end of a life cycle. Because of that, the code coverage for ART was **88%** and for waterfall **49%**, which was calculated by running multiple (100+) test-coverage tests in Visual Studio (through the feature shown in Fig. 9.7), and taking the average percentage for the sample tests. Maintenance costs for ART

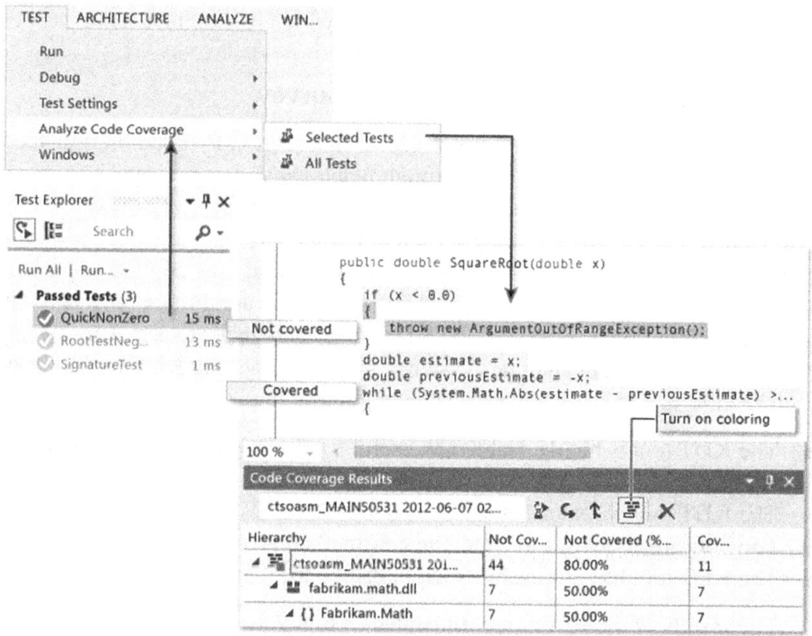

FIGURE 9.7 Code coverage using visual studio (Microsoft's MSDN Visual Studio Developer Manuals, 2016).

TABLE 9.4 Ongoing Maintenance and Validation Costs of ART Versus Waterfall

Cost Type	ART	Waterfall
Unit testing	$5,500	$4,100
Field tests	$1,300	$6,300
Usability testing and refinements	$3,000	$4,100
Posttesting maintenance and development	$4,700	$4,700
Routine monthly validation and verification	$1,200	$3,600
Current ongoing maintenance	$3,000	$5,100
Total	*$17,400*	*$27,900*

and waterfall are shown in Table 9.4. The numbers are rounded up to the 100th dollar, and some of them are best guesses provided by federal employees.

The costs illustrated above are yearly costs, and it shows that ART is an early investment that pays off later in the life of the project, whereas waterfall's costs keep increasing as the project moves forward. This is another major advantage for ART. KDT is also tested through a survey, which is presented next.

4.2 The Agricultural Analyst's Kansei Survey

The experimental survey study presented in this section is based on user feedback gathered using the KDT tool through federal analytics systems. The reason feedback is very important because regular adoption rates of software at the government are usually low, and the process tends to take an extended period of time, especially in the case of our experiment that was applied to agricultural analysts. The KDT tool was used by 80+ federal agricultural analysts. The analyst explored the system and the KDT tool, each analyst spent 1 to 2 weeks browsing through data analytics systems, and then used the KDT tool to insert their feedback. The feedback was categorized into the following groups:

- The KDT tool is highly usable and useful for improving my agricultural data analysis tools
- The KDT tool is somewhat usable and useful
- I do not care about this tool and the new method (not relevant to my work)
- The KDT tool has low usability, and I dislike the method

The results of the survey are illustrated in Fig. 9.8.

As the results show, most users (41%) thought the method is great. Although 13% refused to use it and 17% did not care, 29% of federal employees are

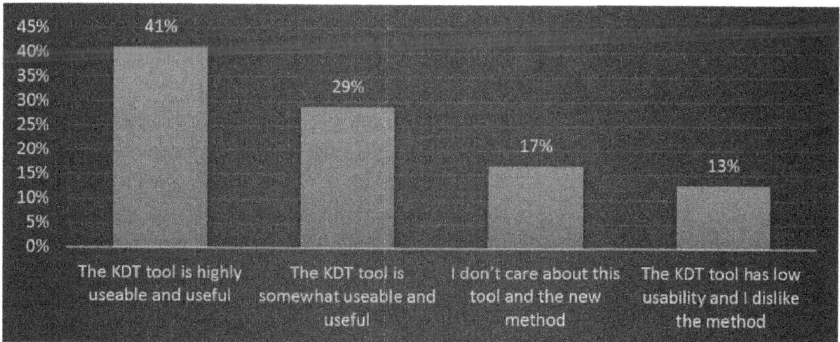

FIGURE 9.8 User satisfaction with Kansei engineering deployment and traceability (KDT).

willing to adopt new methods. Because of the recency of this method and the expected difficulty in adoption at the government, we consider this as a fairly good result.

5. CONCLUSIONS AND FUTURE WORK

As multiple federal agencies are progressively moving into new technologies, the government is contracting with software vendors and universities to deploy new data management and software systems. Efficiency pressures and federal budget constraints are pushing those implementations toward more automation, extra validation challenges, and higher security risks. Federal systems are home for data that support studies for the government in research, legislative work, media requests, measuring performance, and many others. Most importantly, federal analysts struggle with two major aspects of the software life cycle, **testing**, and **adoption**. Based on work done with the US government, this manuscript introduces a comprehensive model for testing and deploying software systems. The life cycle model (FEDAL) provides guidance to developers and engineers throughout the two main phases of deployment (ART and KDT). ART aims to eliminate what the NIST study indicates as the most expensive part of software testing, defining the locations of defects. Based on the experiment, ART was able to accomplish that goal. After ART is deployed, KDT is applied; aiming to collect feedback from the users to improve data systems' adoption.

The future steps of this research include the following: ART needs to be further evaluated on different types of federal software systems, more experimentation needs to be performed, and also, ART needs to be tried with systems that vary in size (the intelligent software system used for the experiment in this chapter is a midsize system). KDT needs to be tested at more agencies to collect more feedback. ART and KDT had fairly successful results, nevertheless, if they eventually facilitate the process of data systems adoption at federal agencies and increase the overall intelligent systems' acceptance by analysts, which would be the main success indicator for this model.

REFERENCES

Afzal, W., Torkar, R., Feldt, R., 2009. A systematic review of search based testing for non-functional system properties. Journal of Information and Software Technology 51.

Almagro, L., Martorell, X., 2012. Statistical methods in Kansei engineering: a case of statistical engineering. In: Proceedings of the International Journal on Quality and Reliability Engineering, pp. 563–573.

Batarseh, F.A., 2012. Incremental Lifecycle Validation of Knowledge-Based Systems through CommonKADS (Ph.D. Dissertation Registered at the University of Central Florida and the Library of Congress).

Batarseh, F.A., Gonzalez, A.J., 2011. Incremental lifecycle validation of knowledge-based systems through CommonKADS. In: IEEE Transactions on System, Man, and Cybernetics, Part A: Systems and Humans.

Batarseh, F.A., Gonzalez, A.J., 2015. Predicting Failures in Contextual Software Development through Data Analytics. Springer's Transactions on Software Quality, pp. 1–18.

Dalal, S., Chhillar, R., 2012. Case studies of the most common and severe types of software system failures. In: Proceedings of the International Journal of Advanced Research in Computer Science and Software Engineering, pp. 341–347.

DePillis, L., 2013. The way government does tech is outdated and risky. Washington Post.

Groth, R., 1999. Industry Applications of Data Mining. Pearson Higher Education Publishers, pp. 191–211 (Chapter 8). ISBN-13: 9780130862716.

Hartono, M., Chua, T.K., 2011. How the Kano model contributes to Kansei engineering in services. Journal of Ergonomics 987–1004.

Introduction to Data Analytics Handbook, 2006. Migrant and Seasonal Head Start Technical Assistance Center, Academy for Educational Development.

I.S.O. Standard 9126, 2016. Available from: http://www.iso.org/iso/home/standards.htm.

Jing, L., 2004. Data Mining Applications in Higher Education (SPSS Executive Report).

Koh, H.C., Gerald, T., 2001. Data mining applications in healthcare. Journal of Healthcare Information Management 19 (2).

McCall, J., Richards, P., Walters, G., 1997. Factors in Software Quality. National Technical Information Service (NTIS). AD-A 049-014-015-055.

Microsoft's MSDN Visual Studio Developer Manuals, 2016. Available at: msdn.microsoft.com.

Nagamachi, M., 1995. Kansei engineering: a new ergonomic consumer-oriented technology for product development. In: Proceedings of the International Journal of Industrial Ergonomics, pp. 3–11.

Pace Systems, 2017. Available from: www.pace-systems.com/.

Shi, F., Xu, J., Sun, S., 2012. Employing Rough and Association Rule Mining in Kansei Knowledge Extraction, vol. 196, pp. 118–128.

Smith, M., Cohen, T., October 2013. A CNN Report on the ACA Website: Problems with Health Website. .

Sommerville, I., 1982. Software Engineering, eighth Ed. Addison-Wesley. ISBN: 9780321313799 (Chapters 1–5).

The National Institute of Standards and Technology (NIST), 2015. Available from: www.nist.gov.

Turing, A.M., 1950. Computing machinery and intelligence. Journal of the Mind 59, 433–460.

United States Government Accountability Office (GAO), 2012. Software Development, Effective Practices and Federal Challenges in Applying Agile Methods. GAO-12-681 gao.gov.

Yang, C.-C., 2011. A classification-based Kansei engineering system for modeling consumers' affective responses and analyzing product form features. Proceedings of Expert Systems with Applications 38 (9), 11382–11393.

Yu Ming-Tun, H., 2012. Designing Software to Shape Open Government Policy (A Dissertation Presented to the Faculty of Princeton University in Candidacy for the Degree of Doctor of Philosophy). The Department of Computer Science, Princeton University.

FURTHER READING

Harmann, M. 2002. The Role of Artificial Intelligence in Software Engineering. Crest Center, University College London.

Schütte, S., 2002. Designing Feelings into Products Integrating Kansei Engineering Methodology in Product Development (A Thesis Published with the Linköping Studies in Science and Technology, Thesis No. 946).

Chapter 10

Federal Big Data Analytics in the Health Domain: An Ontological Approach to Data Interoperability

Erik W. Kuiler, Connie L. McNeely
George Mason University, Arlington, VA, United States

An ontology reflects a negotiated equilibrium between certainty and precision.
Kuiler and McNeely

1. INTRODUCTION

Over the last decade, the health domain has witnessed an extraordinary growth in data-driven medicine as a result of the introduction of, for example, electronic health records (EHRs), digital imaging, digitized procedures, increasing sophistication in laboratory test formulation, the real-time availability of sensor data, and, what stands out in the popular press, the introduction of genomics-related projects (Ohno-Machado, 2012; Shah and Tenenbaum, 2012). Such advances in data generation have informed the discourse on the importance of data exchange and interoperability as well as the broader applicability of the term "big data" to health data analytics (Sahoo et al., 2013). Moreover, big data occupies a critical position on the federal health policy and research agenda, with emphasis on leveraging large, complex data sets to manage population health, drive down disease rates, and control costs.

The scope of the health domain is extensive: patient-focused clinical health, public health, population health, environmental health, global health, and international collaborative research (e.g., genomics and ecological systems). Big health data sets may contain both structured and unstructured data, including clinical device sensor data, reflecting the operationalization of the Internet of Things (IoT), process-driven data, and socioeconomic information. Taking advantage of the availability of big health data sets and adaptable information and communications technology (ICT) capabilities, health care and ICT providers will collaborate

Federal Data Science. http://dx.doi.org/10.1016/B978-0-12-812443-7.00010-7

to support the practice of evidence-based medicine. In 2012, for example, Sloane Kettering Cancer Center and IBM entered into a partnership to use the IBM-developed Watson machine learning system to provide medical personnel the heuristics necessary to identify the most recent evidence-based treatment options and tailor those to meet individual patients' needs (Doyle-Lindrud, 2015). Similarly, in 2016 IBM and Pfizer, Inc., entered into an agreement to use IBM's Watson for Drug Discovery, a new entry in the Watson Health product suite, to discover new drugs to accelerate immunooncology (Bartlett, 2016).

Discussions in the ICT industry and academic journals and in the popular press are typically restricted to technical issues that attend the discovery, dissemination, and analysis of big data sets. However, in the discourse on big data analytics, it is difficult to overlook the burgeoning development of the global IoT, in which devices and humans interact as agents that ingest, transform, and use data in real time to act and perform their tasks as efficaciously and efficiently as possible within predetermined thresholds. However, the complexity attending big data analytics presents multiple challenges. Because massive data sets frequently come from widely distributed sources, tracking data lineage and provenance through their life cycles requires rigorously applied standards, governance, and management processes to ensure consistent quality and secure, authorized data access. In this context, an ontological approach provides an effective means to achieve data interoperability. With the rapid transformation of the health domain, enabled by big data analytics and IoT expansion, the importance of ontologies and their associated lexica cannot be overstated: they function as enablers of data interoperability and information exchanges that transcend national borders, creating international epistemic communities, each of which encapsulates systems of concepts, beliefs, and practices within shared purviews that frame purposive action. Accordingly, critical aspects of ontology-based data interoperability and information exchange are explored here in terms of their applicability to big data analytics in the health domain.

2. DATA INTEROPERABILITY IN THE HEALTH DOMAIN

Data interoperability refers to "the ability of two or more systems or components to exchange data and use information," which "provides many challenges on how to get the information, exchange data, and use the information in understanding it and being able to process it" (European Commission, 2009, p. 9). In reference to the IoT, four aspects of interoperability require specific consideration: technical, syntactical, semantic, and organizational. *Technical interoperability* focuses on the efficacy of the hardware/software system and platform configurations and communications protocols that enable machine-to-machine communications. *Syntactical interoperability* requires the consistent use of data formats (for example, Electronic Data Interchange (EDI) record layouts). Messages transferred by communication protocols must have a well-defined syntax and encoding regimen, even if it is only in the form of bit tables.

However, many protocols carry data or content, which can be represented using high-level syntaxes such as HTML or XML. *Semantic interoperability* supports the human interpretation of the significant meaning of content embedded in a message rather than machine interpretation of that content. *Organizational interoperability* reflects an organization's ability to communicate and transfer data effectively, even though they may be managed by different information systems over different infrastructures and across different geographic regions and cultures. Organizational interoperability depends on the successful integration of technical, syntactical, and semantic interoperability capabilities.

In the United States, these kinds of interoperability are among the primary goals of the Office of the National Coordinator (ONC) for Health Information Technology (HIT) in the Department of Health and Human Services (DHHS). The ONC supports health information exchange (HIE) by formulating applicable policies, services, and standards. The Fast Health Interoperability Resources framework was used to develop the web-based Consolidated Continuity of Care Document (CCCD) and Quality Reporting Document Architecture implementations that comply with such web standards as XML, JSON, and HTTP.[1] Also, the ANSI X12 EDI standard supports the transmission of Healthcare Claim and Claim Payment/Advice data (transactions 835 and 837). The ONC has developed an interoperability road map for the US health domain, including rules of engagement and accountability, secure network infrastructure and transport techniques, consistent authorization, industry-wide testing and certification infrastructure, consistent data semantics, consistent data formats, secure standard services, accurate individual data matching, health care directories, and resource locations. In addition, the ONC emphasizes the development and implementation of health care provider workflows and practices that support consistent data sharing and integration from multiple sources (ONC, 2016).

In the United States, the impetus for health data interoperability came during 2008–10. Under the HIT for Economic and Clinical Health (HITECH) component of the American Recovery and Reinvestment Act of 2009, the Centers for Medicare and Medicaid reimburse health service providers for using electronic documents in formats certified to comply with HITECH's Meaningful Use (MU) standards. The Patient Protection and Affordable Care Act of 2010 promotes access to health care and a greater use of electronically transmitted documentation that is expected to provide a framework for the electronic exchange of health information that complies with all legal requirements and standards.

Of particular importance to data interoperability is the transmission of data that conform to predetermined standards, conventions, and practices that

1. The CCCD is the former Health Information Technology Standards Panel (HITSP) C32 Component that has since been subsumed by, and implemented using, the Health Level 7 (HL7) interoperability framework. HITSP operated under the auspices of DHHS Agency for Healthcare Research and Quality (AHRQ); see http://www.ahrq.gov/. For HL7 International, see http://www.hl7.org/.

are encapsulated in the operations of lexicon and ontology, metadata, and access and security components of the data governance framework. Electronic health documentation overcomes the limitations imposed by paper records—idiosyncratic interpretability, inconsistent formats, and indifferent quality of information—with less likelihood of error in interpretation or legibility. Electronic health documentation may contain both structured and unstructured data (e.g., coded diagnostic data, clinician's notes, personal genomic data, and X-ray images) and usually takes one of three forms, each of which must comply with predetermined standards before they are authorized for use: an electronic medical record for use by authorized personnel within a health care organization, a personal health record (PHR) that provides health care information about an individual from diverse sources (clinicians, care givers, insurance providers, and support groups) for the individual's personal use, and an EHR that provides health-related information about an individual that may be created, managed, and exchanged by authorized clinical personnel (ONC, 2008).

With EHRs, HIE provides the foundation for health data analytics. Adhering to national standards, HIE operationalizes the HITECH MU provisions by enabling the electronic conveyance of heath information among health care organizations, including case management and referral data, clinical results (laboratory, pathology, medication, allergy, and immunization data), clinical summaries (CCD and PHR extracts), images (including radiology reports and scanned documents), free-form text (office notes, discharge notes, emergency room notes), financial data (claims and payments), performance metrics (providers and institutions), and public health data.

Also, the Internet enables both national and international health data interoperability. Genomic and biological research offers a clear illustration of this point. It is an international effort involving a high level of transnational collaboration and data sharing. For example, the database of nucleic acid sequences is maintained by a consortium comprising institutions from the United States, United Kingdom (UK), and Japan. The US National Center for Biotechnology Information maintains the International Nucleotide Sequence Database Collaboration; the UK-based European Bioinformatics Institute maintains the European Nucleotide Archive; the Japanese National Institute of Genetics supports the DNA Data Bank of Japan and the Center for Information Biology. To ensure synchronized coverage, these organizations share information on a regular basis, relying on common metrics and methods to support translational applications. Bioinformatics organizations are developing computational models of disease processes that can prove beneficial not only to the development or modification of clinical diagnostic protocols and interventions, but also for epidemiology and public health.

Because rights to privacy and security constitute a critical area of concern in big data governance and use, translational bioinformatics and health informatics must incorporate stringent anonymization and security controls to safeguard Personally Identifiable Information. In addition, in the United States, many individuals distrust federal agencies, making it difficult to collect data or to develop master patient registries that support health data exchange. Health data interoperability also highlights data ownership and stewardship questions that must be addressed. For example, under what circumstances may data be shared, and with whom; are patients the owners of their health data, or do government agencies, health care providers, product vendors, or payers own those data, or are these organizations only data stewards, with limited rights to those data?

Although opportunities for collaborative health care provisioning and research have increased, institutional barriers to health data interoperability remain. There is an unwillingness to share data, not only by patients but also by providers, payers, and vendors. The US medical community has been relatively slow to adopt the use of EHRs. A number of cultural factors influence this. For example, clinicians want products that meet their specific processes and protocols, which they consider to be unique to their practices (whether or not that is actually the case is not point). In contrast, EHR product vendors want to capture the largest market share possible at the lowest cost; hence, for them, a generic one-size-fits-all solution is the most efficient production model. Furthermore, because it may be in their self-interest and a business advantage not to share data, EHR product vendors treat health data as proprietary assets and maintain competition-focused product-based barriers (such as autarchic standards and processes) to cross-vendor data sharing, with deleterious effects on the rate of hospital participation in HIE (Adler-Milstein et al., 2016).

Systems that provide data interoperability capabilities are expensive, and many small medical practices cannot afford them. Thus the question arises: who pays for them? The federal government initially funded local and state efforts to increase HIE participation of health care providers with the intent of improving patient care efficiency. Unfortunately, government subsidies are frequently not sufficient. Thus, for example, in 2012 there were 119 operational HIE efforts that engaged more than one-third of US providers; in 2014 there were 106 (Adler-Milstein et al., 2016). In addition, a number of federally funded collaborative programs have come to an end. One such program, the Beacon Community Program, established in 2010, is a representative example. The program had mixed success on expanding health care, achieving data interoperability, and applying consistent quality measures. The program's final assessment noted the apparent inability of the private sector to achieve interoperable systems and suggested the need for national leadership to support their creation (ONC, 2015).

All these factors—solipsistic attitudes in the medical community, lack of sustained support, distrust of government agencies, incompatible standards, maintaining market share—hinder data interoperability and information exchange in the US health domain.

3. ONTOLOGIES AS THE BASIS FOR INTEROPERABILITY

An ontology reflects a social construction of reality, defined in the context of a specific epistemic culture as sets of norms, symbols, human interactions, and processes that collectively facilitate the transformation of data into knowledge (Fernández-Breis et al., 2001; Goffman, 1959; Liu et al., 2012). As noted earlier, health data interoperability requires the integration of technical, syntactical, semantic, and organization interoperability. In the context of this paradigm, *data* are given, or admitted, as facts or individual pieces of information that provide a basis for reasoning or inference to support calculation, analysis, or planning. Data provide the instances that expand an ontology so that it constitutes a knowledge base. *Information* is here defined as the creation of meaning as the result of analyzing, calculating, or otherwise exploring data (by aggregation, combination, decomposition, transformation, correlation, mapping, etc.) within the scope delineated by the ontology. *Knowledge* constitutes the understanding gained from analyzing information.

3.1 Lexicon as the Basis for Semantic Congruity

The use of a lexicon is fundamental for developing an ontology because a lexicon provides a controlled vocabulary that contains the terms and their definitions that collectively delimit an epistemic domain, enabling semantic consistency and reducing ambiguity by enforcing rule-based specificity of meaning. The composition of a term in the lexicon comprises its morphemes (root, stem, affixes, etc.), accompanied by a definition that facilitates homonymy and polysemy resolution. An effective lexicon will also support the correlation of synonyms and the semantic (cultural) contexts in which they occur. Furthermore, to support data interoperability, a lexicon should provide mechanisms to support cross-lexicon mapping and reconciliation. For example, in the health domain, the Systematized Nomenclature of Medicine Clinical Terms (SNOMED-CT), maintained by the International Health Terminology Standards Development Organization, is a multilingual lexicon that provides coded clinical terminology extensively used in EHR management. RxNorm, maintained by the US National Institutes of Health National Library of Medicine, provides a common ("normalized") nomenclature for clinical drugs, with links to their equivalents in a number of other drug vocabularies commonly used in pharmacy management and drug interaction software. The Logical Observation Identifiers Names and Codes (LOINC), managed by Indiana University's Regenstrief Institute, provides a standardized lexicon for reporting laboratory results.

3.2 Ontological Dimensions

Ontologies use lexica to delimit their conceptual purviews, and encapsulate the intellectual histories of epistemic communities and support the development of heuristic instruments that sustain data analytics by delineating interdependencies among terminological categories (classes derived from concepts defined in the lexicon) and their properties. Because the relationships between items in an ontology are semantic, linguistic, and syntactic, a number of fundamental tropes guide the analysis, design, and implementation of ontologies: *simile*, to make a comparison where one thing is likened to another; *meronymy*, to indicate a relationship between a part and a whole; *metonymy*, to indicate a substitution of a characteristic, or quality, of a thing for the thing itself. From a HIT perspective, ontologies can be transformed into ICT-enabled constructs that support the transmission, use, and management of shareable data, such as patient, claims, and payment data, that comply with predetermined semantic and syntactic rules. Furthermore, ontologies provide the semantic congruity, consistency, and clarity to support different statistics-based aggregations, correlations, and regressions (Fernández-Breis et al., 2001; Liu et al., 2012).

The development and use of ontologies as intellectual instruments of discovery have a distinguished history. As envisioned by the pre-Socratic philosopher Parmenides (late 6th/early 5th century BCE), the science of ontology is a heuristic pursuit of knowledge and is essentially rhetorical—i.e., social and discursive.[2] Ontologies not only are useful as viable heuristic instruments, but also serve as descriptive or normative instruments, constituting formal, explicit specifications of shared conceptualizations (Gruber, 1993). In computer and information sciences, an ontology defines a set of representational primitives with which one can model a domain of knowledge or discourse. From the perspective of implementing a database application, an ontology can be viewed as a level of abstraction of data models, analogous to hierarchical and relational models, but intended for modeling knowledge about individuals, their attributes, and their relationships to other individuals (Gruber, 2008). An ontology also can be defined as "a shared conceptualization of a domain that is commonly agreed to by all parties" (Dillon et al., 2008, p. 9). Note that these conceptualizations all emphasize the social and discursive aspects of ontology even when, as in computer science, a technological perspective is included.

A more encompassing perspective on ontology consolidates ICT-focused definitions and classical Greek notions and acknowledges the discursive nature of the pursuit of knowledge. In this sense, an ontology is a named perspective defined over a set of categories (or classes) that collectively delimit a domain of knowledge.[3] A *category* is a named perspective defined over a set of properties.

2. Parmenides (2004), *By Being It Is*, Fragment 2: "[2.1] Well then, I will tell you and you who listen, receive my word—what are the only ways of investigation there are to think [2.3] one, on the one hand, [to think] that 'is' and that it is not possible not to be; this is the way of persuasion, since it accompanies the truth; [2.5] another, on the other hand, [to think] that 'is not,' and that it is necessary not to be; I tell you that this path is completely unknowable, since you will not know that which is not (as it is not unknowable, since you will not know that which is not (as it is not possible) or utter it."

3. For a discussion of the different conceptualizations of ontology in philosophy and in computer science, see Zúñiga (2001).

A *property* is an attribute or characteristic common to the set of instances that constitute a category. Take, for example, "cigarette," "smokeless tobacco," and so on, as categories in a hypothetical tobacco product ontology. In this case, length, diameter, mode of ingestion, etc. are specific properties of a class in a tobacco product ontology. That is, properties define the criteria for instance-inclusion in a specific category in the ontology.

Categories, properties, and their interdependencies collectively constitute a semantic network, axiomatically defined, and reflecting, in part, taxonomic reasoning as well as the application of a propositional logic-based grammar, such as a first-order predicate calculus.[4] A taxonomy is a hierarchical classification scheme that constitutes a named, directed acyclic perspective defined over a set of categories.[5] An ontology can be formally specified as a set of axioms using an ontology specification language, or, for Semantic Web applications, using a web-oriented ontology description language (Corcho and Gómez-Perez, 2000; Sowa, 2000; Knublauch, 2004; McGuiness et al., 2002).[6]

3.3 Ontology Development

Ontologies fulfill multiple functions in supporting data interoperability and ICT systems design. As a *descriptive instrument*, an ontology can help establish the boundaries of a particular domain as well as establish a common understanding of that domain. As a *heuristic instrument*, an ontology, when implemented as a software application, can aid data analytics by, for example, displaying a set of category and property selections that allow an analyst to assign a data item to a specific category. As a *normative instrument*, an ontology can support ICT system design—e.g., by guiding database schema design, by formulating the alethic and deontic rules that govern the inclusion or exclusion of data items in a database, and by establishing value-based thresholds to support risk assessments and risk mitigation strategies.

An ontology reflects a negotiated equilibrium between certainty and precision. Vagueness and ambiguity can no more be fully eliminated from analytics or interpretation than they can be eliminated from human discourse. At best, an

---

4. The concept of semantic network has been used in the artificial intelligence community since the late 1980s (Sowa, 1992, 2000).

5. See Tenenberg (1985) for an early overview of principles of taxonomic reasoning and its utility for axiom formulation.

6. For example, Ontolingua, an ontology definition language based on Knowledge Interchange Format (KIF), a propositional language developed in the 1990s, or Object Management Group's (OMG) Web Ontology Language (OWL). A description of Ontolingua is available from http://www.ksl.stanford.edu/software/ontolingua/. KIF reflects a first-order predicate calculus and is designed to support interoperability between knowledge-based systems. Available from http://www.ksl.stanford.edu/knowledge-sharing/kif/. An overview of OMG OWL is available from http://www.w3.org/TR/owl2-overview/.

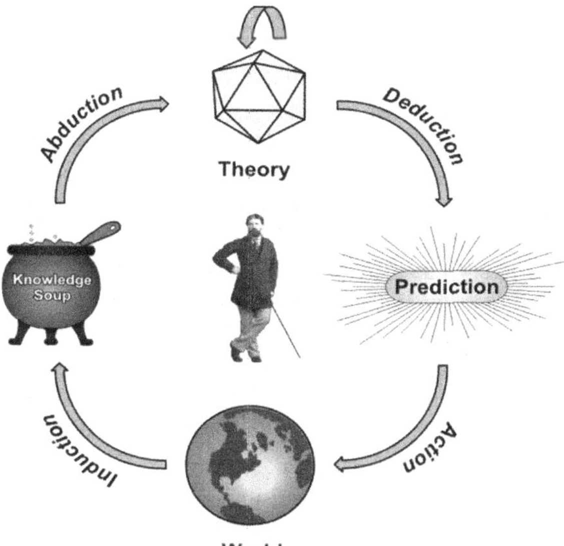

FIGURE 10.1 Ontological development cycle. *Adapted from Sowa, J.F., 2015. The Cognitive Cycle. Available from: http://www.jfsowa.com/pubs/cogcycle.pdf.*

ontology can be reasonably precise and certain, but only within a narrow scope.[7] Thus ontology development comprises a number of interdependent activities that test and correlate theory with "real-world" observations. In practical terms, considerations for developing ontologies include

- large sets of data items with different semantics and formats.
- data from diverse sources (e.g., legacy systems, intraorganizational data sources, and external data sources).
- data with complex relationships (e.g., inheritance, inclusion/exclusion, composition, and containment).
- complex alethic and deontic rules.
- requirements for formal proofs (e.g., contracts or policy enforcement).

As depicted in Fig. 10.1, ontology development operates as a cycle of interdependent activities, which reflect the heuristic nature of ontology development, and may be, but need not be, supported by ICT capabilities. Based on the "cycle of pragmatism" formulated by American philosopher Charles Sanders Peirce, the ontological development cycle integrates reasoning methods with purposeful activities (Peirce, 1997; Sowa, 2015). It offers a meta-level framework

7. Peirce observes: "It is easy to speak with precision upon a general theme. Only, one must commonly surrender all ambition to be certain. It is equally easy to be certain. One has only to be sufficiently vague. It is not so difficult to be pretty precise and fairly certain at once about a very narrow subject." (Peirce, 1958, vol. 4, § 237; cf. Sowa, 2000).

relating methods of reasoning and scientific research in self-correcting cycles (cf. Sowa, 2015).

In the figure, "Knowledge Soup" (stored knowledge) provides one stage-gate in the cycle; "Prediction" provides another. "Theory" and "World" provide the two poles that delimit the domain of an ontology. Abductive reasoning is enthymematic rather than syllogistic and reflects a process of making a set of observations and developing a best explanation of those observations.[8] For example, a nurse observes that a patient in his care appears flushed and drowsy. He looks at the patient's medical history and notices that she had a recent blood transfusion. The nurse, quickly developing an explanation that reflects his training and nursing experience, surmises, or abducts, that the patient may be experiencing an unwanted reaction to the transfusion. To test the likelihood of the correctness of his inference, his supposition, he gathers evidence to support or refute his conjecture (e.g., by talking with the patient and by taking a new set of observations).

Abductive reasoning is fundamental to ICT-focused activities that support knowledge discovery, using, for example, data and text mining paradigms and statistical analytics (cf. Lipscomb, 2012). Deductive reasoning, usually predicated on a syllogistic structure, supports the development of logical inferences (rather than enthymematic inferences), enabled by ICT-implemented inference engines and theorem provers. Inductive reasoning reflects data gathering and developing conclusions from analyzing those data according to existing categories and schemata. Induction plays an important role in developing network and pattern recognition software, for example, to aid a researcher exploring the likelihood of diabetes occurring in a population with certain characteristics. Of course, the activities depicted in Fig. 10.1 collectively support Natural Language Processing (NLP), given that language is paramount not only in communication but also in reasoning. However, in the current state of the discipline, NLP has difficulty coping with linguistic artifacts such as negation and figures of speech (e.g., metaphor, irony, metonymy, or meronymy).[9]

The collaborative reasoning activities depicted in Fig. 10.1 also are applicable to predictive analytics of big health data sets by providing answers to questions more directly related to social practice. For example, what is the likelihood of a health care provider committing fraud, given certain characteristics? As a

---

8. This method approximates an Aristotelian approach that uses an *epagoge* (proof by example derived from experience with similar cases), in which the major premise is true but the minor premise is probable. As a practical application, "an abduction is a hunch that may be a brilliant breakthrough or a dead end. The challenge is to develop a computable method for generating insightful guesses. Beforehand, a guess cannot be justified logically or empirically. Afterwards, its implications can be derived by deduction and be tested against the facts by induction. Peirce's logic of pragmatism integrates the reasoning methods with the twin gates of perception and purposeful action" (Sowa, 2015, p. 3).

9. Pai et al. (2014) provides a useful summary of the current state of NLP and its application to clinical decision support.

corollary, given the historical pattern of the provider's behavior compared with that of other providers, is this one committing fraud? How can fraud detection be enhanced by operationalizing the analytics model?

3.4 Ontology Integration

Many ICT-enabled applications, such as data and text mining and NLP, depend on data interoperability to sustain analytics of data sets from diverse sources consistently and effectively. Frequently, these data, be they structured or unstructured, are formulated to comply with different lexical, grammatical, and syntactic rules; different units of analysis and units of measure; and different semantic frames. For data interoperability to be effective in heterogeneous data environments, a robust data integration framework and mapping methods and techniques are required. To ensure semantic homogeneity, an effective ontology integration framework must resolve a number of difficult, but tractable, heterogeneity issues (Omelayenko, 2001; Wache et al., 2001; Gagnon, 2007):

- *Conceptual modeling heterogeneity*—each data source is specified as a distinct data model, which may be presented visually as, for example, an entity relationship diagram or an object role model.
- *Syntactic heterogeneity*—each data source has its own schema (record, file) layout, predicated on the requirements of the data management software used by the source data provider.
- *Semantic heterogeneity (polysemy)*—each data source reflects a specific lexicon, denotational context, and, possibly, semiotics system.
- *Pragmatic heterogeneity*—each data source has a specific usage context.

There are a number of approaches that support ontology integration. In a single ontology approach, a global lexicon and ontology ensure semantic consistency across multiple, disparate data sources, provided that these data sources share the same domain perspective. Any differences in lexica or epistemic perspectives increase the level of difficulty of ontology integration. In a multiple ontology approach, cross-ontology mapping is a standard method for ensuring semantic constituency. Multiple ontology integration can be labor-intensive, however. In a hybrid approach, ontology integration depends on the implementation of a single lexicon for use by multiple local ontologies (Gruber, 1995; Wache et al., 2002). Another approach to ontology integration is to manipulate the sets of axioms that constitute the formal specification of the ontologies. In this case, integration might be achieved by, for example, creating a more general theory by deleting axioms (contraction), or by making the theory more specific by adding axioms, or by renaming one or more axiomatic predicates (Woods, 1991; Sowa, 2000).

Multiple ontologies also can be integrated by focusing on categories, properties, and semantic network relationships. One method is to merge categories based on the union of their set of properties. Another method is to create a superclass over multiple classes with the properties that these classes have in common. In this approach, the classes retain only their nonshareable properties. If there are class name conflicts (homonyms), names can be changed to resolve the conflict. To remove ambiguity, it may be necessary to create multiple disjoint categories, possibly with a supercategory to accommodate shared properties.

3.5 Ontology Operationalization

To support data interoperability, ontologies can ensure semantic consistency and provide specifications for database design. Under most circumstances, ontology categories can be transformed to database table specifications or file record layouts, as discussed earlier. Doing so involves the development of interoperationable database schemata based on specifying relevant entities, attributes, integrity constraints, and organizational rules. Ontology properties may be mapped to database table or record attributes or fields. Ontology relationships correlate with inter- and intrarecord navigation indices, such as referential integrity constraints (primary key-foreign key constraints in RDBMS-supported databases) and cardinality constraints (1:1, 1:M, M:M). Axioms are, perhaps, the most difficult ontology components to operationalize because they may be realized as file records, fields, optionality (NULL, NOT NULL) specifications, and integrity constraints, usually in the form of an organization's operational rules.

In a health domain that encompasses IoT and big data sets, many organizations have had to develop new rules and algorithms to integrate, standardize, and manage data while preserving patient privacy and supporting information security. Specific rules that reflect an organization's operational environment must indicate the requirements that data must meet to be useful rather than be excluded as "noise." Thus there are *alethic rules* that indicate possibility or impossibility of the existence of an entity (logical necessity). For example, in a clinical database, it is necessary for a patient to have an identification number; without such a number, the patient cannot formally exist in that environment. *Deontic rules* are normative and constrain the formal existence of an item. Such rules come into play, for example, in determining the propriety of an insurance claim as either valid or invalid and the attendant penalties for filing an invalid claim. *Category assignment rules* specify the restrictions imposed on categorical membership, the values that may be assigned to data fields (for example, a restriction on a range of values that may be selected for a field), and the valid interdependencies that may exist between data items. For example, "AK—Alaska," "MD—Maryland," and "MI—Michigan" are members of a hypothetical US state reference code set. In addition, the application of category assignment rules reflects the use of controlled sets of codes, their values, and their interpretations that delimit the range and domain of data fields. *Taxonomic assignment rules* stipulate branch points in a database schema, based on a particular value of a data attribute, or values of a set of data attributes, reflecting an "if…then…else" logic.

3.6 Metadata Foundations

Metadata are required to support data interoperability and the transformation of an ontology into useable information resources. Metadata refer to any data that describe the container as well as the contents of a collection of data. In this context, containers include record layouts as well as data set specifications. Consider here the Dublin Core Metadata Initiative (DCMI), which supports a widely used metadata standard for identifying, cataloging, and managing data sets and their contents (cf. Dublin Core, 2017).[10] DCMI provides a set of metadata elements for developing descriptive records of data and information resources: subject, description, type, source, title, relation, coverage, creator, publisher, contributor, rights, date, format, identifier, language, audience, provenance, rights holder, instructional method, accrual method, accrual periodicity, and accrual policy (Dublin Core, 2005).

Dublin Core metadata are interoperable, and, to enhance Semantic Web applicability, DCMI focuses on four interoperability levels (in order from the least to most complex): (1) *shared term definitions*, shared vocabularies defined in natural language; (2) *formal semantic interoperability*, shared vocabularies based on formal semantics; (3) *description set syntactic interoperability*, shared formal vocabularies in exchangeable records; and (4) *description set profile interoperability*, shared formal vocabularies and constraints in records (Dublin Core, 2009). In addition, DCMI provides guidance to translate metadata records into OMG Resource Description Framework statements. ICT-supported registries are important tools for managing metadata and industry standards that address metadata registry implementation include ISO/IEC 11179, *Specification and Standardization of Data Elements*, and ANSI X3.285, *Metamodel for the Management of Shareable Data.*

4. CONCLUSION

Ontologies and their associated lexica encapsulate the cultural norms and concepts, their properties, and interdependencies that collectively define the health domain and provide the foundations for data interoperability. Effective policy

---

10. The Dublin Core Metadata Element Set is the result of discussions held at a 1995 workshop sponsored by Online Computer Library Center (with its provenance in the Ohio College Library Center) and the National Center for Supercomputing Applications. Because the workshop was held in Dublin, Ohio, the element set was named the Dublin Core. The Dublin Core Metadata Initiative (DCMI) supports innovation in metadata design and best practices across the metadata environment. DCMI offers a research infrastructure and supports conferences, meetings, and workshops, and educational efforts to promote widespread acceptance of metadata standards and best practices (cf. Dublin Core, 2017).

formulation in the health domain is critical for providing the institutional guidance, program development, and implementation necessary to achieve societal well-being. Furthermore, facilitating the effective use of integrated and, where appropriate, synthesized data is fundamental to support program execution and oversight. Accordingly, data interoperability is critical in the formulation and dissemination of information and practice in the health domain, and an ontology-based approach offers a practical and adaptable means for addressing a number of related challenges presented in the construction and analytics of massive health data sets. The design and operationalization of ontologies and lexica provide the means to ensure semantic congruity and syntactic conformance, facilitating the incorporation, synthesis, and integration of data from multiple sources. When properly implemented, an ontology provides the basis for transforming big data into information and, subsequently, into knowledge that may be exchanged among health care stakeholders—researchers, providers, payers, and patients—with tools for addressing both cultural and technological heterogeneities that hinder HIE. This point emphasizes the critical need for national leadership and policy makers to create and foster an environment conducive to sustainable HIE and data interoperability.

REFERENCES

Adler-Milstein, J., Lin, S., Jha, A., 2016. The number of health information exchange efforts is declining, leaving the viability of broad clinical data exchange uncertain. Health Affairs 35 (7), 1278–1285.

Bartlett, J., 2016. Pfizer to use IBM Watson supercomputing to find new drugs. Boston Business Journal. Available from: http://www.bizjournals.com/boston/news/2016/12/01/pfizer-to-use-ibm-watson-supercomputing-to-find.html?surround=etf&u=Bs06NVDBFTkxq1E71lhgaQ01342d42&t=1487168404&j=77272891.

Corcho, O., Gómez-Perez, A., 2000. A roadmap to ontology specification languages. In: Dieng, R., Corby, O. (Eds.), Knowledge Engineering and Knowledge Management Methods, Models, and Tools: Proceedings of the 12th International Conference on EKAW 2000, Juan-les-Pins, France, October 2–6, 2000. Springer Publishing Company, New York.

Dillon, T., Chang, E., Hadzic, M., Wongthontham, P., 2008. Differentiating conceptual modelling from data modelling, knowledge modelling, and ontology modelling and a notation for ontology modelling. Paper Presented at the Fifth Asia-Pacific Conference on Conceptual Modelling, Australian Computing Society, Wollongong, New South Wales, Australia, January 1, 2008.

Doyle-Lindrud, S., 2015. Watson will see you now: a supercomputer to help clinicians make informed treatment decisions. Clinical Journal of Oncology Nursing 19 (1), 31–32.

Dublin Core, 2005. Using Dublin Core: The Elements. Available from: http://dublincore.org/documents/usageguide/elements.shtml#maincontent.

Dublin Core, 2009. Interoperability Levels for Dublin Core Metadata. Available from: http://dublincore.org/documents/interoperability-levels/.

Dublin Core, 2017. Metadata Initiative. Available from: http://dublincore.org/.

European Commission, European Research Cluster on the Internet of Things (IERC), 2009. The Internet of Things: IoT Governance, Privacy and Security Issues. Available from: http://www.internet-of-things-research.eu/pdf/IERC_Position_Paper_IoT_Governance_Privacy_Security_Final.pdf.

Fernández-Breis, J.T., Valencia-García, R., Martínez-Béjar, R., Cantos-Gómez, P., 2001. A context-driven approach for knowledge acquisition: application to a leukemia domain. In: Akman, V., Bouquet, P., Thomason, R., Young, R., (Eds.), Modeling and Using Context, CONTEXT 2001. Lecture Notes in Computer Science, vol. 2116. Springer, Heidelberg.

Gagnon, M., 2007. Ontology-based integration of data sources. In: 10th International Conference on Information Fusion, Quebec Canada, July 9–12, 2007.

Goffman, E., 1959. The Presentation of Self in Everyday Life. Doubleday, New York.

Gruber, T.R., 1993. A translation approach to portable ontology specifications. Knowledge Acquisition 5, 199–220.

Gruber, T., 1995. Towards principles for the design of ontologies for knowledge sharing. International Journal of Human Computer Studies 43 (5/6), 907–928.

Gruber, T.R., 2008. Ontology. In: Lui, L., Tamer Özsu, M. (Eds.), Encyclopedia of Database Systems. Springer-Verlag, New York, NY. Available from: http://tomgruber.org/writing/ontology-definition-2007.htm.

Knublauch, H., 2004. Ontology-Driven Software Development in the Context of the Semantic Web – An Example Scenario with Protégé and OWL. Available from: http://people.cs.ksu.edu/~abreed/CIS890/References/ontologydrivensoftwaredevelopment.pdf.

Lipscomb, M., 2012. Abductive reasoning and qualitative research. Nursing Philosophy 13, 244–256.

Liu, Y., Coulet, A., LePendu, P., Shah, N.H., 2012. Using ontology-based annotation to profile disease research. Journal of the American Medical Informatics Association 19, e177–e186.

McGuinness, D.L., Fikes, R., Hendler, J., Stein, L.A., 2002. DAML + OIL: an ontology language for the semantic web. Intelligent Systems 772–780.

Office of the National Coordinator (ONC), 2008. The National Alliance for Health Information Technology Report to the National Coordinator for Health Information Technology on Defining Key Health Information Technology Terms, Health Information Technology. Available from: http://www.hitechanswers.net/wp-content/uploads/2013/05/NAHIT-Definitions2008.pdf.

Office of the National Coordinator (ONC), 2015. Evaluation of the Beacon Community Cooperative Agreement Program. Health Information Technology/NORC at the University of Chicago. Available from: https://www.healthit.gov/sites/default/files/norc_beacon_evaluation_final_report_final.pdf.

Office of the National Coordinator (ONC), 2016. Connecting Health and Care for the Nation: A Shared Nationwide Interoperability Roadmap, Health Information Technology. Available from: https://www.healthit.gov/sites/default/files/hie-interoperability/nationwide-interoperability-roadmap-final-version-1.0.pdf.

Ohno-Machado, L., 2012. Big science, big data, and the big role for biomedical informatics. Journal of the American Medical Informatics Association 19 (e1).

Omelayenko, B., 2001. Syntactic-level ontology integration rules for e-commerce. In: Proceedings of the FLAIRS-01. American Association for Artificial Intelligence, pp. 1–5. Available from: http://www.aaai.org/Papers/FLAIRS/2001/FLAIRS01-063.pdf.

Pai, V.M., Rodgers, M., Conroy, E., Luo, J., Zhou, R., Seto, B., 2014. Workshop on using natural language processing for enhanced clinical decision making: an executive summary. Journal of the American Medical Informatics Association 21, 1–4.

Parmenides, 2004. In: Cordero, N.L. (Ed.), By Being It Is: The Thesis of Parmenides. Parmenides Publishing, Las Vegas, NV.

Peirce, C.S., 1958. In: Hartshorne, C., Wiess, P., Burks, A. (Eds.). Hartshorne, C., Wiess, P., Burks, A. (Eds.), Collected Papers of C.S. Peirce, 8 vols. Harvard University Press, Cambridge, MA.

Peirce, C.S., 1997. Pragmatism as a principle and method of right thinking. In: Turrisi, P.A. (Ed.), The 1903 Lectures on Pragmatism. SUNY Press, Albany, NY.

Sahoo, S., Jayapandian, C., Garg, G., Kaffashi, F., Chung, S., Bozorgi, A., Chen, C., Loparo, K., Lhatoo, S., Zhang, G., 2013. Heart beats in the cloud: distributed analysis of electrophysiological 'big data' using cloud computing for epilepsy clinical research. Journal of the American Medical Informatics Association. http://dx.doi.org/10.1136/amiajnl-2013-002156.

Shah, N.H., Tenebaum, J.D., 2012. The coming of age of data-driven medicine: translational bioinformatics' next frontier. Journal of the American Medical Informatics Association 19, e1–e2.

Sowa, J.F., 1992. Semantic Networks. Available from: http://www.jfsowa.com/pubs/semnet.pdf.

Sowa, J.F., 2000. Knowledge Representation: Logical, Philosophical, and Computational Foundations. Brooks/Cole, Pacific Grove, CA.

Sowa, J.F., 2015. The Cognitive Cycle. Available from: http://www.jfsowa.com/pubs/cogcycle.pdf.

Tenenberg, J.D., 1985. Taxonomic reasoning. In: Proceeding of IJCA '85 9th International Joint Conference on Artificial Intelligence, vol. 1, pp. 191–193.

Wache, H., Vögele, T., Visser, U., Stuckenschmidt, H., Schuster, G., Neumann, H., Hübner, S., 2001. Ontology-based integration of information – a survey of existing approaches. In: Proceedings of IJCAI-01 Workshop: Ontologies and Information Sharing, WA: Seattle, August 4–5, 2001, pp. 108–117.

Wache, H., Visser, U., Scholz, T., 2002. Ontology construction: an iterative and dynamic task. In: Proceedings, Florida Artificial Intelligence Research Society Conference, Pensacola, Florida, USA, pp. 445–449.

Woods, W.A., 1991. Understanding subsumption and taxonomy: a framework for progress. In: Sowa, J.F. (Ed.), Principle of Semantic Networks: Explorations in the Representation of Knowledge. Morgan Kaufman Publishers, Inc., San Mateo, CA, pp. 45–94.

Zúñiga, G.L., 2001. Ontology: its transformation from philosophy to information systems. In: Proceedings of the International Conference on Formal Ontology in Information Systems, Ogunquit, ME, October 17–19, 2001, vol. 2001, pp. 187–197.

Chapter 11

Geospatial Data Discovery, Management, and Analysis at National Aeronautics and Space Administration (NASA)

Manzhu Yu, Min Sun
George Mason University, Fairfax, VA, United States

The road ahead will be long. Our climb will be steep

Barack Obama

1. INTRODUCTION

During the past century, our capabilities have been significantly improved to observe and record various physical parameters of the Earth's surface at a speed of terabytes to petabytes per day. This improvement is mainly due to the invention of digital computers for information processing, the launch of satellites, and the evolution of remote sensing technologies (Christian, 2005). Meanwhile, the capability of large-scale scientific simulation has been enhanced by leveraging the increasingly powerful computing resources and producing massive amounts of data. For example, tens of gigabytes of data are generated routinely by climate and weather forecasting models, such as the Goddard Institute for Space Studies General Circulation Model ModelE (Schmidt et al., 2014), and Weather Research and Forecasting model (Michalakes et al., 2005). The next generation of global cloud resolving simulations will produce terabytes of data on a per-time step basis (Kouzes et al., 2009).

These big observational and simulation data are of great value to different domains with Geosciences because they provide data about the Earth at specific time snapshots (Yang et al., 2011). Various types of applications and scientific studies can benefit from these data, such as natural disaster monitoring, early warning, and response system (Basher, 2006); climate (atmospheric, land, and ocean) modeling (Kiehl et al., 1998); and aiding policy making (Goetz, 2007).

Federal Data Science. http://dx.doi.org/10.1016/B978-0-12-812443-7.00011-9

With a wider and larger demand for these observational and simulation data from the research and application community, government agencies (e.g., National Aeronautics and Space Administration (NASA), National Oceanic and Atmospheric Administration) and intergovernmental groups (e.g., the Group on Earth Observation) started to build systems to help store, manage, access, and share these data sets to address worldwide and regional problems (Lautenbacher, 2005).

Owing to the great value of big geospatial and spatiotemporal data, research efforts have been made to address their characteristics regarding the volume, variety, velocity, and veracity (Big Data's 4 Vs). The nature of these data, including heterogeneous, spatiotemporal, and multidimensional, however, are severely behind the rate of data growth and deserves further attention. Also, it is important to study the types of research questions that these big geospatial and spatiotemporal data bring forth. In this chapter, we will focus on investigating the different research questions for better resource discovery, access, storage, and management of data, and algorithms or tools for performing large-scale simulation and analysis. For each research question, we proceed to briefly review the associated challenges, and demonstrate with sample works applied to address these challenges. We conclude the chapter with a review and future directions.

2. GEOSPATIAL DATA DISCOVERY

To better disseminate, analyze, and utilize geospatial data, a variety of centralized catalogs and portals have been developed, such as the Geospatial One-Stop Portal,[1] OpenTopography,[2] and Unidata.[3] Through these cyberinfrastructures (CIs), data can be discovered and accessed by users from relevant disciplines, agencies, and counties. These data discovery CIs are playing significant roles in globalization, which requires sharing and coordinating Earth observation and simulation data sets across jurisdiction boundaries to enable global geoscience research and emergency response (NRC, 2003). However, geospatial data discovery is challenging because, in addition to sheer volume, they can be globally distributed and heterogeneous, representing different Earth phenomena using different conceptualization, representations, units, and formats. Therefore metadata and semantics for describing the resources are central to this integration process to enable users to discover and evaluate the information resources that exist for Earth science studies and applications.

Metadata, usually archived in catalogs, have been developed to describe the data sets better and provide searching capabilities for domain users to discover data sets needed (Huang et al., 2011). For the purpose of interoperability, i.e.,

1. https://catalog.data.gov/dataset.
2. http://www.opentopography.org/.
3. http://www.unidata.ucar.edu/.

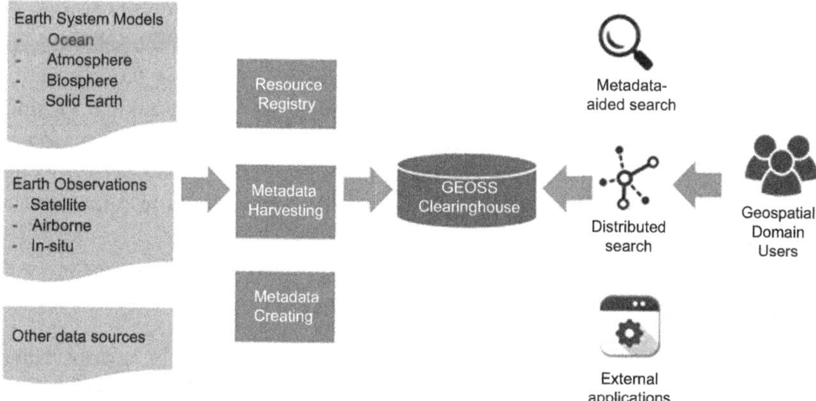

FIGURE 11.1 Global Earth Observation System of Systems Clearinghouse metadata registry workflow.

metadata sharing across different archive centers, the interfaces of metadata need to be standardized based on service-oriented architecture. As an exemplifying case study, Global Earth Observation System of Systems (GEOSS) Clearinghouse is developed to provide big metadata covering services from various domains and promote the use of registered services for different applications (Liu et al., 2011). To overcome the structural and semantic heterogeneity of geospatial data, GEOSS Clearinghouse integrates a series of open, community consensus standards for metadata harvesting and user access. The metadata of the distributed data resources are registered into the GEOSS Clearinghouse as a data and service container (Fig. 11.1), ensuring the consistency of the metadata and providing multiple ways for remote resource discovery (Li et al., 2010).

However, when metadata are shared across different domains, it becomes difficult to discover the right data because of the mismatch in semantics between end users and the metadata content (Li et al., 2008). Therefore ontology needs to be integrated with metadata to capture the meaning of user query and the potential usage of data in a knowledge management and utilization fashion to increase the possible searching results and to refine final results. One sample work of integrating semantics into geospatial data discovery is the GeoSearch system (Gui et al., 2013). GeoSearch utilizes semantic reasoning to enrich the search query and filter search results by relevance, thus improving the search accuracy. For example, if a user searches "water wms," a total of 42 WMS (Web Map Service) records can be found, and within these records, water-related terminologies, such as "ice," "ocean," "lake," and "river," are also included in the search. In contrast, without semantic assistance, the resulting number of records will be 985 but the majority of the search results are irrelevant, with only 22 WMS records, and the others are not WMS but data sets. Therefore by collaborating with the semantic supporting system, users can get more of what they really need and eliminate those irrelevant results.

Based on GEOSS Clearinghouse and GeoSearch, MUDROD (Jiang et al., 2016) integrates metadata, ontology, and user logs to improve the search accuracy of geospatial data. SWEET ontology[4] is used to assist in the estimation of semantic query similarity and the improvement of search ranking. Besides metadata and ontology, user logs are mined to discover user behavior regarding their search history, clickstream, and data popularity. These user behaviors are valuable information for data mining, and further provide more relevant data ranking and suggestion.

3. BIG GEOSPATIAL DATA MANAGEMENT

Geospatial data are growing at tremendous speed, collected in various ways, including photogrammetry and remote sensing, and more recently through laser scanning, mobile mapping, geo-located, sensors, geo-tagged web contents, volunteer geographic information, and simulations (Li et al., 2016a,b). Part of this growth is due to the increasing powerful computation capabilities to handle the big geospatial data. For example, in climate science, climate simulations are conducted in an increasingly higher spatiotemporal resolution, which largely relies on the capabilities of parallel computing. However, the higher spatiotemporal resolution simulation has brought researchers new challenge, which is to deal with the efficient management, analysis, and visualizations of the simulation output. To efficiently store, manage, and query the big geospatial data set, one must consider the data structure, modeling, and indexing for a case-dependent, customized big geospatial data management solution.

In climate science, NetCDF/HDF is one of the most commonly used data structures, which consists of multidimensional variables within their coordinate systems and some of their named auxiliary attributes (Rew and Davis, 1990). However, the classic data model has two obvious limitations: (1) lack of support for nested structure, ragged arrays, unsigned data types, and user defined types; and (2) limited scalability due to the flat name space for dimensions and variables.

To address the limitations of the classic data model, Li et al. (2017) proposed an improved data model based on NetCDF/HDF, which included to contain additional named variables, dimensions, attributes, groups, and types. The variables can be divided into different groups by a certain characteristic, such as the model groups that generated these data. When storing these variables in a physical file, each two-dimensional grid will be decomposed into a one-dimensional byte stream and stored separately, one by one, in a data file.

With this improved data model, efficient management of these data is still challenging because of the large data volume, as well as the intrinsic high-dimensional nature of geoscience data. In addition, distributed file systems, such as MapReduce, have become increasingly popular in managing and processing

4. https://sweet.jpl.nasa.gov/.

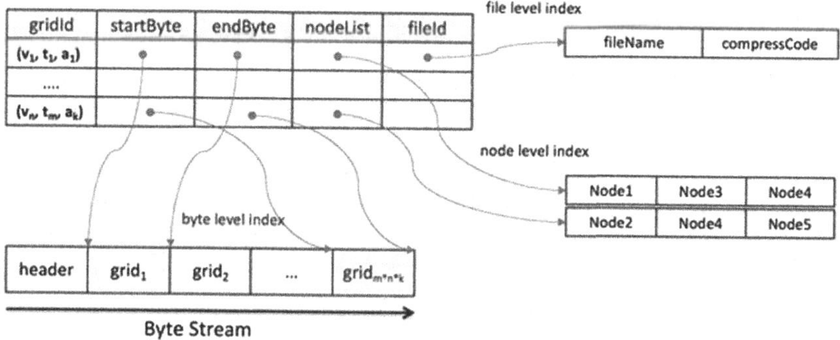

FIGURE 11.2 The structure of spatiotemporal index.

big data, but there is a gap between the MapReduce model and the efficient handling of the traditional climate data format (NetCDF/HDF). To tackle this challenge, Li et al. (2017) also proposed a spatiotemporal indexing approach to efficiently manage and process big climate data with MapReduce in a highly scalable environment. Using this approach, big climate data are directly stored in a Hadoop Distributed File System in its original, native file format. A spatiotemporal index is built to bridge the logical array-based data model and the physical data layout, which enables fast data retrieval when performing spatiotemporal queries (Fig. 11.2). Based on the index, a data-partitioning algorithm is applied to enable MapReduce to achieve high data locality, as well as balance the workload. The proposed indexing approach is evaluated using the NASA Modern-Era Retrospective Analysis for Research and Applications (MERRA) climate reanalysis data set. The experimental results show that the index can significantly accelerate querying and processing (~10× speedup compared with the baseline test using the same computing cluster), while keeping the index-to-data ratio small (0.0328%). The applicability of the indexing approach is demonstrated by a climate anomaly detection deployed on a NASA Hadoop cluster. This approach is also able to support efficient processing of general array-based spatiotemporal data in various geoscience domains without special configuration on a Hadoop cluster.

4. LARGE-SCALE SCIENTIFIC SIMULATION

The analysis and prediction of climate change and natural disasters rely extensively on efficient processing of big geospatial data. The processing efficiency is challenging to achieve since geospatial data volume is growing tremendously, and the data capturing the information of natural phenomena are intrinsically complex and spatiotemporal. Scientists and researchers usually utilize high-performance computing to reduce the computing time for large-scale scientific simulation. High-performance computing can harness high-performance hardware such as multi–central processing unit (CPU)-core computers or clusters to enhance the computing capacity and reduce the execution time for scientific models. Parallel processing, which is embedded in high-performance computing, partitions the model domain into subdomains so that each subdomain can be processed on a distinct CPU core, thus reducing the total processing time. During parallel processing, distributed-memory parallel computers need to communicate and synchronize so that the variable calculation is consistent for the entire model domain. However, communication and synchronization costs a great amount of time; thus intelligently decomposing the model domain and allocating computing tasks to the right computing resource can greatly improve simulation performance.

4.1 Spatiotemporal Thinking to Optimize High-Performance Computing

To improve computational efficiency, the spatiotemporal pattern of natural phenomena can be considered into decomposition and scheduling process. Taking dust simulation as an example, three types of spatiotemporal characteristics of dust events can be leveraged: (1) dust event is a clustered phenomenon with local dust concentration clustering; (2) dust event is space–time continuity, meaning it initiates, moves, evolves, and slows down in a continuous fashion; (3) dust events are isolated or restricted in their spatiotemporal scope and can, therefore, be treated event by event.

Based on the first spatiotemporal pattern of dust event, clustered phenomena, dust storm simulation needs deal with clustered regions, and the scheduling method can take advantage of this characteristic. For example, if two computing nodes and eight subdomains are determined, then the first, third, fifth, and seventh subdomains will be dispatched to the first computing node and the remaining subdomains will be dispatched to the second computing node. The default decomposition and scheduling method uses the nonneighbor method, illustrated in Fig. 11.3B. In the subdomain and computing nodes experiment, two computing nodes are utilized: half continuous subdomains are dispatched on the first computing node and the rest are dispatched on the other computing node in a

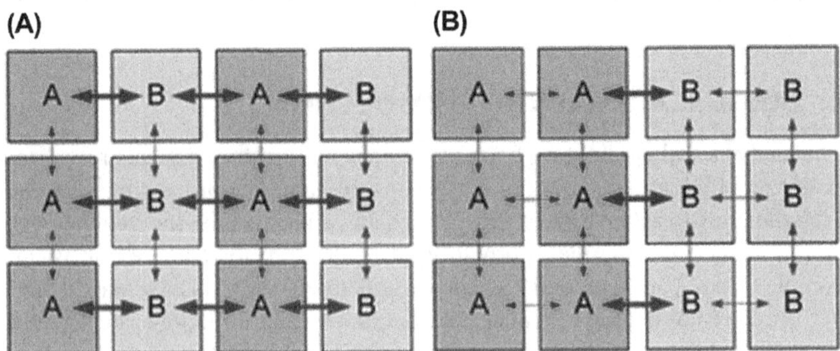

FIGURE 11.3 Two scheduling methods for dispatching 12 subdomains to two computing nodes, A and B (Huang et al., 2013a). (A) Uncluster allocation and (B) cluster allocation.

neighbor scheduling fashion. In Fig. 11.3, communication costs between different computing nodes are illustrated by thicker arrows, whereas communications within the same computing nodes are illustrated by narrow arrows. Experiments showed that performance improvement factors of approximately 20% on average could be achieved by using the neighbor scheduling method. This result suggests that it is better to dispatch neighbor subdomains to the same computing node to reduce the communication over computer networks.

The second characteristic, space–time continuity, is also associated with data exchange among neighboring subdomains (Yang et al., 2011). Based on the clustering experiment, different clustering methods also result in different computing times. One of the reasons is that data exchange is caused by interactions of physical dynamics along the space, which is not necessarily consistent all over the world. For example, wind velocities are relatively large near the poles and are much smaller in the North–South (meridional) direction than those in the East–West (zonal) direction (Nanjundiah, 1998). Therefore communication needs differ among processors in the South–North (S–N) direction from those of the West–East (W–E) direction. In addition, different domain sizes along the W–E and S–N directions result in different numbers of grid cells along these two directions. Thus for the same degree of parallelization, different decompositions can result in a different communication overhead.

Considering the third characteristics, isolated dust events, high-resolution model simulation can be conducted event by event. Huang et al. (2013a) proposed an adaptive, loosely coupled strategy, which couples a high-resolution/small-scale dust model with a coarse-resolution/large-scale model. Specifically, the adaptive, loosely coupled strategy would (1) run the low-resolution model; (2) identify subdomains of high predicted dust concentrations (Fig. 11.4); and (3) run the higher-resolution model for only those subdomains with much smaller area in parallel. In this approach, the

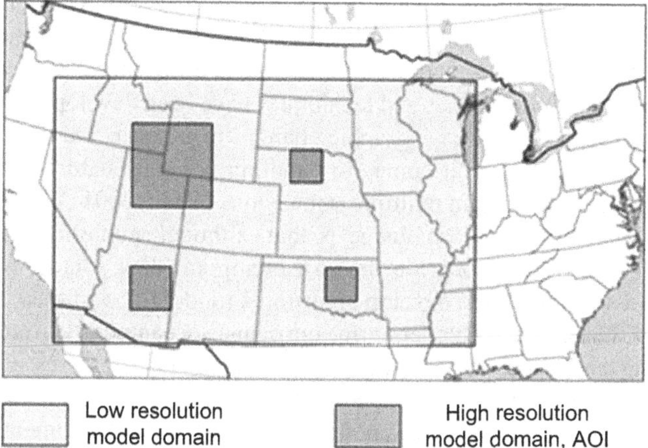

FIGURE 11.4 Low-resolution model domain area and subregions (area of interest, AOI) identified for high-resolution model execution (Huang et al., 2013a).

high-resolution model results for specific subdomains of interest could be obtained more rapidly than an execution of a high-resolution model over the entire domain.

4.2 Cloud Computing to Support Large-Scale Scientific Simulation

One of the problems with traditional computing infrastructure is that it does not scale well with the rapidly increasing data volume and computation intensity. Cloud computing is such a computing platform in that its computing capabilities can be rapidly and elastically provisioned. Cloud computing allows one to have a high-performance cluster in a very short time to process large-scale scientific simulations (Huang et al., 2013b). Existing efforts have been utilizing Hadoop as the platform to perform big geospatial data processing. For example, Li et al. (2016a,b) investigated the approach of adjusting computing resources to efficiently balance workloads, which can automatically scale and allocate the proper amount of cloud resources based on dynamic workload. Such an auto-scaling approach provides a valuable reference to optimize the performance of geospatial applications to address data- and computational-intensity challenges in GIScience in a more cost-efficient manner.

5. SPATIOTEMPORAL DATA MODELING AND ANALYSIS

Natural phenomena, such as those studied in meteorology, oceanography, and geosciences, are intrinsically spatiotemporal (3D: latitude, longitude, and time, or 4D: latitude, longitude, vertical level, and time) in nature and highly dynamic (Yang et al., 2011). To obtain a better understanding of the natural phenomena, it is crucial for scientists and researchers to capture the complex physical processes and evolution patterns, and further improve the capability of simulation and forecasting models. The obtained knowledge or insights may include "where and when natural phenomena happen," "how long a natural phenomenon lasts," or "what the common transport pathway is for a natural phenomenon."

GIScience methodologies and techniques have been developed to assist the understanding of dynamic geographic changes over space and time, but challenges still remain when it comes to handling complex natural phenomena, especially when data are in multiple dimensions (Yuan, 2001; Worboys, 2005; Pultar et al., 2010). One challenge is that, although numerical simulations and earth observations provide the spatiotemporal data source, researchers and scientists still need to develop algorithms to identify and track the movement of features (e.g., thunderstorm, hurricane, ocean eddy). Automatically identifying and tracking features is difficult, because features are moving with changing boundaries and capable of splitting and merging. Another challenge is to formalize the representation of phenomena evolvement to model the event, which can be applied to detect and analyze instances of the event of interest. Data modeling for complex natural phenomena has not been fully implemented in current spatiotemporal data models.

5.1 Spatiotemporal Data Model

Various spatiotemporal data models have been proposed since the 1980s, including the space–time composite (Langran and Chrisman, 1988), the spatiobitemporal model (Worboys, 1994), the three-domain model (Yuan, 1999), and the event-based spatiotemporal model (Peuquet and Duan, 1995), to assist the understanding of dynamic phenomena changes over space and time. Comparing and summarizing the characteristics of the above-mentioned data models, Yuan (1996) discussed their applicability to different kinds of spatiotemporal analysis and concluded with the challenges of using these data models to represent complex spatiotemporal changes. Among the forms of spatiotemporal changes, the most complicated one is a moving field with changing boundaries, capable of splitting and merging, and in need of maintaining a spatiotemporal structure (which may vary over space and time) with other entities or dynamics in a large environment (Yuan, 2009). Most natural phenomena fall into this category, and data modeling for them has not been fully implemented in current spatiotemporal data models (Wang, 2014). Considering the importance of understanding the dynamic physical process of natural phenomena, a spatiotemporal data model that can handle the representation of complex spatiotemporal changes would be of great value to support the analysis and mining of complex processes.

The authors propose a spatiotemporal data model that handles the complexity of natural phenomena in four dimensions. In this research, an event is defined as an occurrence, whereas a process is defined as a sequence of dynamically related static objects that show how the process evolves in space and time. An event consists of one or multiple processes, and these processes interact with each other in one way or another, such as splitting or merging. A process is a single track, which consists of multiple static objects from consecutive time steps. An event has its properties, such as beginning time, ending time, and average coverage, but how the natural phenomenon initiates, transits, develops, and evolves is described by the process. Therefore the characterization and understanding of a dust storm event relies on the analysis of underlying processes within the event. The notion of process and event is consistent with Yuan (2001), Yuan and Hornsby (2007), and Galton and Mizoguchi (2009). Based on this set of notions, the proposed data model consists of four entities: object, interaction, process, and event.

5.2 Tracking Changes and Interactions

Based on the data model, a feature identification and tracking algorithm is developed to reconstruct the four types of entities. Fig. 11.5 shows the

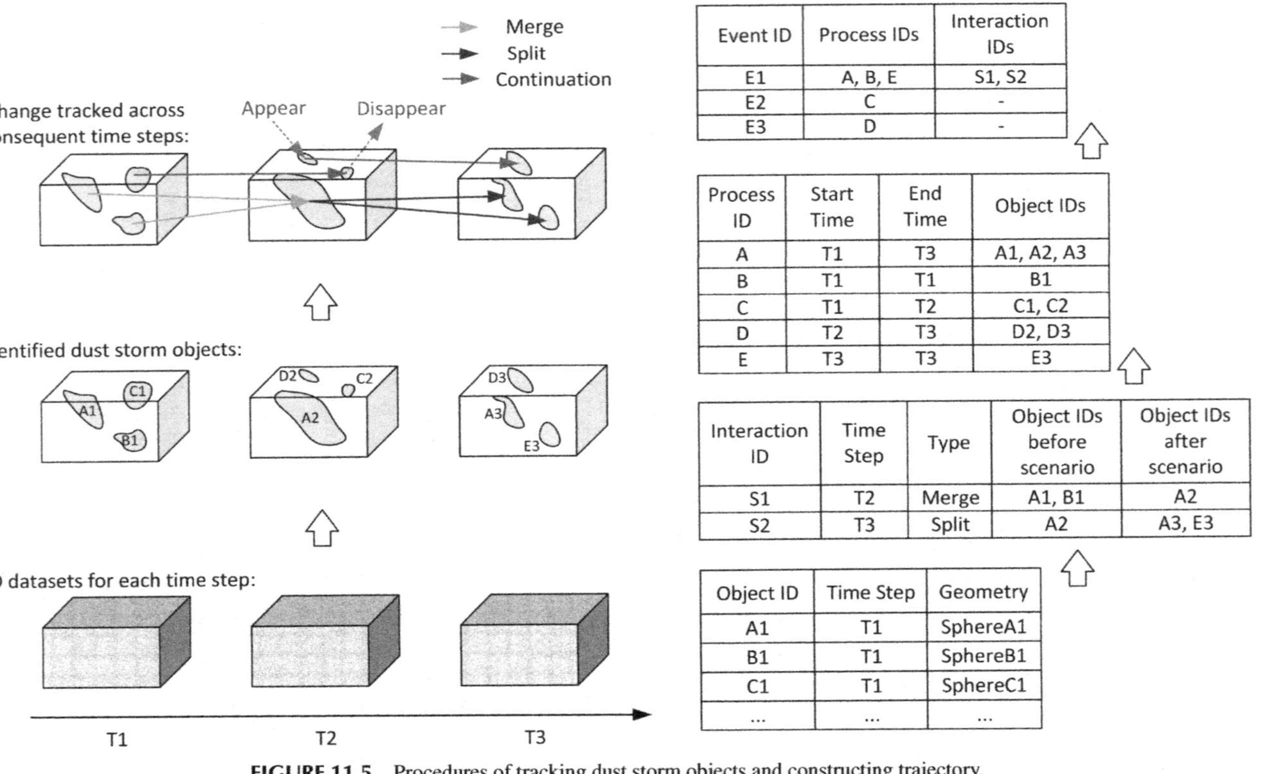

Event ID	Process IDs	Interaction IDs
E1	A, B, E	S1, S2
E2	C	-
E3	D	-

Process ID	Start Time	End Time	Object IDs
A	T1	T3	A1, A2, A3
B	T1	T1	B1
C	T1	T2	C1, C2
D	T2	T3	D2, D3
E	T3	T3	E3

Interaction ID	Time Step	Type	Object IDs before scenario	Object IDs after scenario
S1	T2	Merge	A1, B1	A2
S2	T3	Split	A2	A3, E3

Object ID	Time Step	Geometry
A1	T1	SphereA1
B1	T1	SphereB1
C1	T1	SphereC1
...	...	...

FIGURE 11.5 Procedures of tracking dust storm objects and constructing trajectory.

procedure of constructing the four entities of the data model, which include: (1) identifying dust storm objects; (2) identifying the interactions of objects; (3) tracking the dust storm process; and finally (4) building dust storm events.

In this research, dust events are chosen as a case study to illustrate how this data model can be used to represent and analyze the dynamic movements of phenomena. For an individual dust event, it is essential to understand the physical process of dust uplifting from arid and semiarid regions, transporting in the air, and depositing back to the ground. 4D simulation outputs from a dust model, BSC-DREAM8bv2.0 (Basart et al., 2012), are utilized as the experiment data set. The simulated dust concentration data through 12 months, from December 2013 to November 2014, are used, on a spatial resolution of 0.3 degrees × 0.3 degrees for the broad northern Africa and Europe domain (64.3W–0.76E, 25.7–59.3N). The temporal resolution of the data is hourly. Each time step contains a voxel number of 256 × 196 × 24 (latitude, longitude, pressure level). The identification of dust storm objects at each time step is conducted by a 3D region-grow-based algorithm (Yu and Yang, 2016). This method essentially integrates the idea of region-grow algorithm (Zucker, 1976) into 3D context. In this method, a dust storm object is specified as a contiguous volume with a dust concentration value larger than a value threshold, while its volume is larger than a volume threshold. The movement tracking of dust events is conducted by an overlapping mechanism, which investigates the overlap from features extracted from t with all of the objects extracted from $t+1$. There are several possible cases for the linkage between t and $t+1$, including continuation, merge, split, appear, and disappear. Merging and splitting are constructed as interactions. With the aforementioned five types of linkage, individual dust processes can be constructed and recorded in the process attributed table.

5.3 Spatiotemporal Analytics

The proposed data model represents dust storm events and processes as individual data objects; hence their properties and relationships are readily computable in a geographic information system (GIS). For example, for dust events, it is important to analyze the relationship between dust storm pathways and its originating dust source area. Fig. 11.6A and C show an example of querying the dust events occurring during June 2014 that originate from Libya desert. The pathways of each dust process are displayed in blue, and split and merge interactions are displayed in red and green correspondingly. There are totally 15 events consisting of 72 processes during this time period occurring in the study area. Fig. 11.6B and D illustrates all dust events occurring during June 2014 without constraining on dust source area, which contains 65 events consisting of 335 processes in the entire study area.

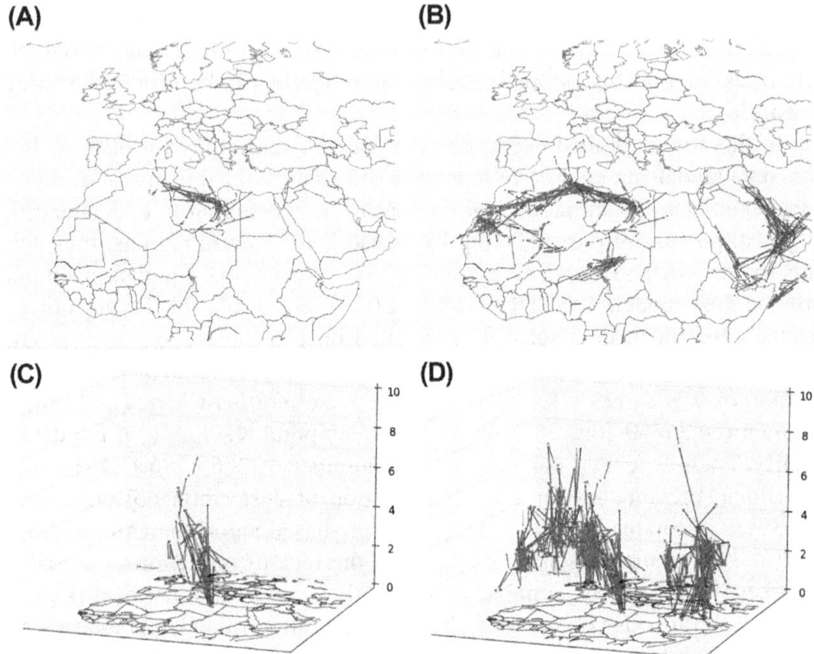

FIGURE 11.6 Example of querying the dust events occurring during June 2014 that originate from Libya desert (A), and without constraining dust source. (B) Dust storm events occurred during June 2014 in study area, (C) 3D view of (A), and (D) 3D view of (B).

6. CONCLUSION AND FUTURE DIRECTIONS

We presented a review of some of the unique challenges of geospatial and spatiotemporal discovery, management, processing, and analytics. We encourage interested readers to refer to the references for further reading.

We have seen tremendous growth in data-driven research and applications in geospatial-related fields during the last two decades. This rapid growth was fueled by the Internet's capability of data production, access, and sharing. The growth of data brings forth a wide range of challenging real-world research questions, and this chapter covers only four of them, including geospatial data discovery, big geospatial data management, large-scale scientific simulation, and spatiotemporal data modeling and analysis. Traditional technologies are weak in addressing these challenges, and increasing research efforts have been established to advance our problem-solving capabilities.

There are also great opportunities for future research and development. First is the intelligent utilization of computing resources. In particular, cloud computing has been proved to be a promising option to handle the complex scalability issues of the big geospatial data discovery, management, and processing (Yang et al., 2011). Second is to fully integrate spatiotemporal concepts into the cutting-edge geospatial research frontier. This challenge requires geoscientists, engineers, and educators from multiple domains to collaborate to solve fundamental problems, such as how to formalize the representation of highly dynamic natural phenomena (e.g., tsunami or hurricane) and how to produce high-resolution weather prediction with broad geographic coverage for regional emergency responses.

Answers to some of the above-mentioned challenges will emerge over time as we continue to develop and testify new applications to big geospatial and spatiotemporal data. Significant breakthrough might require diligent collaborations with other disciplines, such as computer science, atmospheric science, and ocean science.

ACKNOWLEDGMENTS

Research presented in this chapter was funded by National Aeronautics and Space Administration (NASA), Federal Geographic Data Committee (FGDC), National Science Foundation (NSF), and US Geological Survey (USGS).

REFERENCES

Basart, S., Pérez, C., Nickovic, S., Cuevas, E., Baldasano, J.M., 2012. Development and evaluation of the BSC-DREAM8b dust regional model over Northern Africa, the Mediterranean and the Middle East. Tellus B 64.

Basher, R., 2006. Global early warning systems for natural hazards: systematic and people-centred. Philosophical Transactions of the Royal Society of London A: Mathematical, Physical and Engineering Sciences 364 (1845), 2167–2182.

Christian, E., 2005. Planning for the global earth observation system of systems (GEOSS). Space Policy 21 (2), 105–109.

Galton, A., Mizoguchi, R., 2009. The water falls but the waterfall does not fall: new perspectives on objects, processes and events. Applied Ontology 4 (2), 71–107.

Goetz, S., 2007. Crisis in earth observation. Science 315 (5820), 1767.

Gui, Z., Yang, C., Xia, J., Liu, K., Xu, C., Li, J., Lostritto, P., 2013. A performance, semantic and service quality-enhanced distributed search engine for improving geospatial resource discovery. International Journal of Geographical Information Science 27 (6), 1109–1132.

Huang, M., Maidment, D.R., Tian, Y., 2011. Using SOA and RIAs for water data discovery and retrieval. Environmental Modelling and Software 26 (11), 1309–1324.

Huang, Q., Yang, C., Benedict, K., Rezgui, A., Xie, J., Xia, J., Chen, S., 2013a. Using adaptively coupled models and high-performance computing for enabling the computability of dust storm forecasting. International Journal of Geographical Information Science 27 (4), 765–784.

Huang, Q., Yang, C., Benedict, K., Chen, S., Rezgui, A., Xie, J., 2013b. Utilize cloud computing to support dust storm forecasting. International Journal of Digital Earth 6 (4), 338–355.

Jiang, Y., Li, Y., Yang, C., Armstrong, E.M., Huang, T., Moroni, D., 2016. Reconstructing sessions from data discovery and access logs to build a semantic knowledge base for improving data discovery. ISPRS International Journal of Geo-information 5 (5), 54.

Kiehl, J., Hack, J., Bonan, G., Boville, B., Williamson, D., Rasch, P., 1998. The national center for atmospheric research community climate model: CCM3. Journal of Climate 11 (6), 1131–1149.

Kouzes, R.T., Anderson, G.A., Elbert, S.T., Gorton, I., Gracio, D.K., 2009. The changing paradigm of data-intensive computing. Computer 42 (1), 26–34.

Langran, G., Chrisman, N.R., 1988. A framework for temporal geographic information. Cartographica: The International Journal for Geographic Information and Geovisualization 25 (3), 1–14.

Lautenbacher, C., 2005. The global earth observation system of systems (GEOSS). In: Local to Global Data Interoperability-Challenges and Technologies. IEEE, pp. 47–50.

Li, W., Yang, C., Raskin, R., 2008. A semantic enhanced search for spatial web portals. In: AAAI Spring Symposium: Semantic Scientific Knowledge Integration, pp. 47–50.

Li, W., Yang, C., Yang, C., 2010. An active crawler for discovering geospatial web services and their distribution pattern–a case study of OGC Web Map Service. International Journal of Geographical Information Science 24 (8), 1127–1147.

Li, S., Dragicevic, S., Castro, F.A., Sester, M., Winter, S., Coltekin, A., Pettit, C., Jiang, B., Haworth, J., Stein, A., 2016a. Geospatial big data handling theory and methods: a review and research challenges. ISPRS Journal of Photogrammetry and Remote Sensing 115, 119–133.

Li, Z., Yang, C., Liu, K., Hu, F., Jin, B., 2016b. Automatic scaling Hadoop in the cloud for efficient process of big geospatial data. ISPRS International Journal of Geo-Information 5 (10), 173.

Li, Z., Hu, F., Schnase, J.L., Duffy, D.Q., Lee, T., Bowen, M.K., Yang, C., 2017. A spatiotemporal indexing approach for efficient processing of big array-based climate data with MapReduce. International Journal of Geographical Information Science 31 (1), 17–35.

Liu, K., Yang, C., Li, W., Li, Z., Wu, H., Rezgui, A., Xia, J., 2011. The GEOSS clearinghouse high performance search engine. In: IEEE – 2011 19th International Conference on Geoinformatics, pp. 1–4.

Michalakes, J., Dudhia, J., Gill, D., Henderson, T., Klemp, J., Skamarock, W., Wang, W., 2005. The weather research and forecast model: software architecture and performance. In: Proceedings of the Eleventh ECMWF Workshop on the Use of High Performance Computing in Meteorology. World Scientific, Singapore, pp. 156–168.

Nanjundiah, R.S., 1998. Strategies for parallel implementation of a global spectral atmospheric general circulation model. In: IEEE – HIPC'98. 5th International Conference on High Performance Computing, pp. 452–458.

National Research Council (NRC), 2003. Living on an Active Earth: Perspectives on Earthquake Science. National Academies Press.

Peuquet, D.J., Duan, N., 1995. An event-based spatiotemporal data model (ESTDM) for temporal analysis of geographical data. International Journal of Geographical Information Systems 9 (1), 7–24.

Pultar, E., Cova, T.J., Yuan, M., Goodchild, M.F., 2010. EDGIS: a dynamic GIS based on space time points. International Journal of Geographical Information Science 24 (3), 329–346.

Rew, R., Davis, G., 1990. NetCDF: an interface for scientific data access. IEEE Computer Graphics and Applications 10 (4), 76–82.

Schmidt, G.A., Kelley, M., Nazarenko, L., Ruedy, R., Russell, G.L., Aleinov, I., Bauer, M., Bauer, S.E., Bhat, M.K., Bleck, R., 2014. Configuration and assessment of the GISS ModelE2 contributions to the CMIP5 archive. Journal of Advances in Modeling Earth Systems 6 (1), 141–184.

Wang, Y., 2014. MeteoInfo: GIS software for meteorological data visualization and analysis. Meteorological Applications 21 (2), 360–368.

Worboys, M.F., 1994. A unified model for spatial and temporal information. The Computer Journal 37 (1), 26–34.

Worboys, M., 2005. Event-oriented approaches to geographic phenomena. International Journal of Geographical Information Science 19 (1), 1–28.

Yang, C., Goodchild, M., Huang, Q., Nebert, D., Raskin, R., Xu, Y., Bambacus, M., Fay, D., 2011. Spatial cloud computing: how can the geospatial sciences use and help shape cloud computing? International Journal of Digital Earth 4 (4), 305–329.

Yu, M., Yang, C., 2016. A 3D multi-threshold, region-growing algorithm for identifying dust storm features from model simulations. International Journal of Geographical Information Science 1–23.

Yuan, M., 1996. Temporal GIS and spatio-temporal modeling. In: Proceedings of Third International Conference Workshop on Integrating GIS and Environment Modeling, Santa Fe, NM.

Yuan, M., 1999. Use of a three-domain representation to enhance GIS support for complex spatiotemporal queries. Transactions in GIS 3 (2), 137–159.

Yuan, M., 2001. Representing complex geographic phenomena in GIS. Cartography and Geographic Information Science 28 (2), 83–96.

Yuan, M., 2009. 13 toward knowledge discovery about geographic dynamics in spatiotemporal databases. In: Geographic Data Mining and Knowledge Discovery, p. 347.

Yuan, M., Hornsby, K.S., 2007. Computation and Visualization for Understanding Dynamics in Geographic Domains: A Research Agenda. CRC Press.

Zucker, S.W., 1976. Region growing: childhood and adolescence. Computer Graphics and Image Processing 5 (3), 382–399.

Chapter 12

Intelligent Automation Tools and Software Engines for Managing Federal Agricultural Data

Feras A. Batarseh, Gowtham Ramamoorthy, Manish Dashora, Ruixin Yang
George Mason University, Fairfax, VA, United States

Program testing can be used to show the presence of bugs, but never to show their absence.

Edsger Dijkstra

1. INTRODUCTION AND MOTIVATION

Federal analysts and agricultural economists face many data-related problems in completing their daily tasks. The work done by economists is driven by numbers, doing calculations, validating data, and generating formulas. Sometimes analysts are presented with an easy question, but getting the answer can be quite difficult because of the lack of a clean and clear data infrastructure (Bertota et al., 2014). For example, a journalist calls a federal agricultural analyst and asks the following: what is the sum of tons of potatoes that were exported to the countries of south Asia in the years 2013–17? This may seem a very simple question if the numbers are available, but answering such a question without the correct data, accurate aggregations, and a direct access to valid numbers could be a daunting task. In the 1980s and 1990s, most analysts used trivial software tools for data management, such as excel sheets. That is considered to be a time-consuming task that requires much maintenance, accompanied with data sharing and transparency difficulties. Sharing big amounts of data in excel sheets, for instance, can be very risky. A small mistake in data can make a noticeable difference in federal calculations, which eventually may lead to invalid results (and therefore affect government credibility). What if an analyst wants to know the total production of carrots in the United States, in million pounds, from last 5 years? There are hundreds of such questions that

Federal Data Science. http://dx.doi.org/10.1016/B978-0-12-812443-7.00012-0

economical analysts face daily (Joseph and Johnson, 2013). In addition, there can be scenarios where an analyst wants to refer to data that are historical or many years old (from the 1930s or 1920s). In such cases, the analyst has to search for nonelectronic forms, such as books, papers, and old documents, which becomes almost impossible to manage and maintain. Such questions, however, using data science tools, can be answered in few seconds by just querying a data warehouse. In addition, managing calculations can be really tough; how can analysts run calculations automatically given all the different types of inputs, time, and geographical dimensions? How can they validate data and the results? Many questions require a comprehensive and fully automated data system. This chapter aims to present a solution.

2. RELATED WORK

For every federal agency, user and system requirements can constantly change, and hence the data models related to the requirements have to be modified accordingly. For example, when a system needs to be upgraded (replacing federal legacy systems) (USDA, 2016), or two units are merged (two divisions within a federal agency), data validation is required. Actions such as transferring data between servers and applications' transformations are possibly accompanied with reorganizations and consolidations that can exacerbate the quality of data.

2.1 Data Validation Methods

Federal agencies share data all the time, but as it can be imagined, if "moving" data are not sufficiently validated, it would be extremely risky to the owners of the data in the future (federal analysts). Data validation is concerned with not only comparing data between the source and the destination, but also ensuring the *semantical correctness*, *completeness*, and *consistency*, and the *interoperability* of data sets (Haller et al., 2011; Spivak, 2012; Thalheim and Wang, 2013). Especially nowadays, simple databases have been growing tremendously into "big data." Many business intelligence and decision-making tasks rely highly on data; lower data quality may result in unsuccessful strategies, which will further bring severe losses to a federal agency (public trust, tax payers' conflicts, media wars, accountability, and many other issues). Therefore migrating data successfully could be a difficult pursuit, and it should be planned, implemented, and validated thoroughly. To mitigate the risks of data anomalies, the research community proposed different state-of-the-art models and solutions. In his paper, Fine studies the relation between validation and dependability (Fine et al., 2003). Flamos et al. (2010) proposed a data validation platform, called *Scientific Reference System Scorecard*, that assesses and monitors data quality. Others (Haller et al., 2011) did not only propose a generic model for data validation but also introduced a set of testing and quality assurance measures to address the risks of validation during data migration, which is relevant to the work presented here. Their paper categorized data migration risks into three levels: the business level, the information technology (IT) management level, and the data level. For each level, the authors listed associated risks in detail. For example, risks at the level of IT management include data, information loss, or target information stability. Another key contribution of that paper is that the authors described different testing techniques used in data migration, such as migration run tests, appearance tests, processability tests, and integration tests, and then mapped them with corresponding risks (Haller et al., 2011). However, the paper did not address the harmonization issue (merging data from multiple sources, which is a very commonplace use case at the government). Similar articles also mentioned common errors and risks of data migration that should be taken into consideration, but did not provide any direct validation solutions that could be applied to government (Anavi-Chaput et al., 2000). Different organizations may have different requirements regarding how to evaluate data quality, and there is not a single universal technique that can be applied across the board. To that point, one paper proposed a hybrid approach (consists of multiple methods) to dynamically assess the data quality and validate it (Woodall et al., 2013).

As for the industry, Oracle, for example, provides Advanced Customer Support Services specifically for data migration (Oracle, 2016). Their services include different ways of transporting data, as well as comprehensive validation and migration testing. For validating data in Oracle, as described in Anavi-Chaput et al. (2000) and Oracle (2016), up to three full test migrations may be executed, to identify potential problems during the process. Oracle claims that it can save a huge amount of time and effort, and lower the risks by using its services. However, using such tools can create dependency by federal agency on the software vendor, something that is preferably avoided. Therefore building an in-house validation engine is recommended.

2.2 Data Security and Integrity Methods

Besides validation, data security is another major worry for any federal project. The goal is to prevent malicious or unauthorized users from accessing or modifying data stored in databases. Similar to generic security concept, three key aspects determine how secured a data warehouse is: *Confidentiality*, *Integrity*, and *Availability*. *Confidentiality* refers to disclosing data only to authorized users (this is addressed in the tool proposed later in this chapter). To verify a user's identity and control access to data, database management systems use different methods of authentication. For example, Oracle provides multiple ways of authentication, such as authentication by Operating System, by network, by Oracle database, and by secure socket layer protocol (Oracle, 2016). *Integrity* refers to protecting the database from unauthorized writing. Data stored in databases should not be modified improperly. *Availability* means legitimate users should be able to access the database anytime they are allowed to. Firewalls are widely used to protect databases from Denial of Service attacks; that ensures the continuous availability of the database. In the age of big data, protecting data is one of the most critical missions. As some agencies make their databases available online for public access, breaches of databases inevitably occur (domestic or international).

For decades, the research society has been devoting much effort on finding effective methodologies regarding data security, from access control, encryption, to privacy protection. Access control was one of the earliest database security measures proposed and widely used. In most cases, access control models can be categorized into three classes: discretionary access control (DAC), mandatory access control, and role-based access control. Oracle database, for example, provides comprehensive DAC by regulating all user access to named objects through privileges. Encryption is also a popular methodology to enforce confidentiality of databases. For commercial products, Oracle database uses Transparent Data Encryption that is able to encrypt data on storage media transparent (by storing encryption keys outside the database) (Scheier, 2015; Klazema, 2016; Batarseh et al., 2017). The engines presented in this chapter validate, secure, calculate, and automate agricultural data. The federal tool and the two engines (the main contribution of this chapter) are presented next.

3. THE INTELLIGENT FEDERAL MATH ENGINE

As it is well known, government agencies depend on many mathematical formulas for their reports and forecasts. For instance, in agriculture economics at the US Department of Agriculture (USDA), vegetables imports, crops exports, total supply, per capita use, and many other calculations require detailed formulas and a flexible engine that can intelligently capture the math and execute it. Performing basic arithmetic operations is deemed to be a straightforward process (addition, subtraction, division, and multiplication). Nevertheless, what if the federal analyst is asked to do so for million data records in a matter of minutes? That cannot be accomplished without automation. Automation engines can help the economic analyst execute math, and obtain the results fast. The math engine presented in this section allows the analyst to input formulas, define inputs and outputs, and get results. For example, a formula involved in calculating the total exports of fresh carrots, in pounds, from 1990 to 2006, in the United States, requires the addition and multiplication of multiple numbers. That process requires an intelligent engine that can contain all the knowledge of performing such math. The math engine has been developed to run on top of an agricultural data warehouse; the technical details are presented next.

3.1 Inputs, Outputs, and Process of the Math Engine

The math engine is a set of stored procedures, which has two in-built functions and various conditional loops for different time frames; the main technology driving the engine is *Dynamic SQL* (SQL, 2016). The math engine calculates numerous mathematical formulas and stores the output results in a database. Most government agencies have a centralized warehouse in servers where all (or most of) the data reside. The math engine was designed to require three tables from a database: (1) Math Logic Table, where the formulas reside; (2) Variable Outputs Table, where the outputs are *pushed* to; and the (3) Data Inputs Table, where the inputs are *pulled* from by the engine. The math engine first reads each formula listed in the *Math logic* table, and then it extracts data from *Data inputs* before it saves the results in the *Variable output*. The engine does that by replacing the actual values in the formula using a *recursive* method. The output results are either inserted or updated (if they exist prior) in the Variables output table.

For instance, a formula may involve converting data in kilograms (kg) to pounds (lb) or aggregate a statistical type of any agricultural commodity (i.e., a formula is used to convert carrot production in kg to lb or add total fresh carrot production for a particular time frame). The main contents of such formula include: "Inputs, time dimensions, and geography dimensions." An example formula would look like this:

(Production average (lettuce) + Production average (carrots) + Production average (onions))/1000

Inputs: *Production (lettuce), Production (carrots), Production (onions)*
Input/output geographical dimension: USA
Input time dimension value: all months
Output time dimension value: all years

The input time dimension value describes what type of data is to be extracted from the data inputs table, for example, in the above-mentioned sample formula the input time dimension value is *all months*, so the engine picks only the *monthly* data as input to the formula. Similarly, the output time dimension value describes the type of output data to be generated. From the earlier example, there is output time dimension value of *all years*, so the engine converts the monthly data to yearly data and inserts the yearly data into the output table. The common time dimensions considered in the math engine are **years**, **seasons**, **quarters**, **months**, **weeks**, and **days**.

The math engine has *seven* major steps for execution:

1. Declaring variables and reading the contents of the formula
2. Reading time dimensions of a formula
3. Extracting values based on time dimension values
4. Replacing the variables with actual values in the formula using recursive technique
5. Executing the formula using dynamic SQL
6. Insert or update process of output results
7. Exception handling

The next section introduces the seven-step process in greater detail.

3.2 The Seven-Step Math Process

The first step is **declaring variables and reading the contents of the formula**: In any stored procedure, the first step would be declaring variables. The math engine uses three types of variables, loop counting variables, storage variables, and time dimension variables. Millions of formulas are stored in the math logic table and the engine has to have means of managing the execution order. The engine then has certain variables declared for temporarily storing the contents of a complicated part of a formula that is used somewhere else afterward, and these variables are called storage variables. Lastly, the most important attribute in an agricultural math input is "time," based on which the engine operates and generates output records.

Reading time dimensions of a formula: Business formulas are bound to time, i.e., every formula would generate output records that are specific to a certain time frame. For example, "all months" is one of the time dimension values used in the code, it tells the engine to extract all the monthly data for the inputs used in the formula. This way, all the time dimensions, such as "Years," "Quarters," "Weeks," and "Seasons," can be analyzed by the engine.

Extracting values based on time dimension values: After reading all the contents in the formula, the engine understands which data should be extracted. SQL is used to filter through the values and to query the correct numbers.

Replacing the variables with actual values in the formula using the recursive technique: After extracting the actual values for the formula, a function replaces formula input strings with actual values (numbers). To achieve that, the engine relies on the recursive method. Consider the sample formula presented in the previous section; the input strings and the variables are replaced with actual numbers. Table 12.1 shows how the source data (with actual values) are organized and then Table 12.2 shows how it looks after it is queried and injected into the formula.

TABLE 12.1 Components of a Formula

Formula	Inputs Involved	Actual Value	Time Dimension
(Production (onions) + imports (onions))/10,000	Production (onions)	1245 million pounds	2011
(Production (onions) + imports (onions))/10,000	Imports (onions)	1085 million pounds	2011
(Production (onions) + imports (onions))/10,000	Production (onions)	1500 million pounds	2012
(Production (onions) + imports (onions))/10,000	Imports (onions)	980 million pounds	2012

TABLE 12.2 How the Intelligent Math Engine Performs the Math

ID	Formula	Time Dimension	Output Name
1	(1245 + 1085)/10,000	2011	Total supply, 2011
2	(1500 + 980)/10,000	2012	Total supply, 2012

Executing the formula: Consider the simple formula: (1245 + 1085)/10,000, which needs to be calculated in SQL. A built-in stored procedure called ***sp_executesql*** is used to perform the math. This stored procedure (along with dynamic SQL statements) can perform math and the output value is then stored in a variable. The engine uses the following code to perform math for the recursive output values:

```
exec sp_executesql 'SELECT @value_out=(1245+1085) / 10000','@value_out decimal(15,3) OUTPUT', @value_out=@value output
```

Insert or update process of output results: The intelligent math engine is executed whenever the analyst wants to execute it (through the tool that is presented in a later section). The engine should perform insert operation for new math executions, because it generates new output values, or it should perform update operation for existing business rule executions. All of the above-mentioned steps handle errors through *catching exceptions*; the intelligent engines presented in this chapter have rigorous validation methods; that is presented next.

4. VALIDATION AND VERIFICATION OF FEDERAL AGRICULTURAL DATA

In data management, *validating data* is an important stage (if not the most important). In general, validation is achieved in two steps. First, before the data enters the database, and second, after data manipulation processes (through the math engine, for example). By executing the math engine, new data are stored in the database; therefore the validation engine checks whether the data generated are valid. The output results of the engine could be pushed to an excel file (so the nontechnical federal analyst can review and evaluate it)—all that is performed through the tool that is presented later in this chapter. The validation engine is a combination of three SQL stored procedures: *Flag validation*, *Technical validation*, and *Summary statistics validation*. These procedures validate data existing at different time frames; all three procedures would validate data at different time dimensions: monthly, quarterly, and yearly. Flag validation is a stored procedure that assigns flags/colors to data points, and identifies statistical outliers. The flag engine picks the current time frame's data point (imports in 2017, for example) and compares it with the flag rule; then based on the existing user-defined flag range a data point is given a specific flag color. The flag

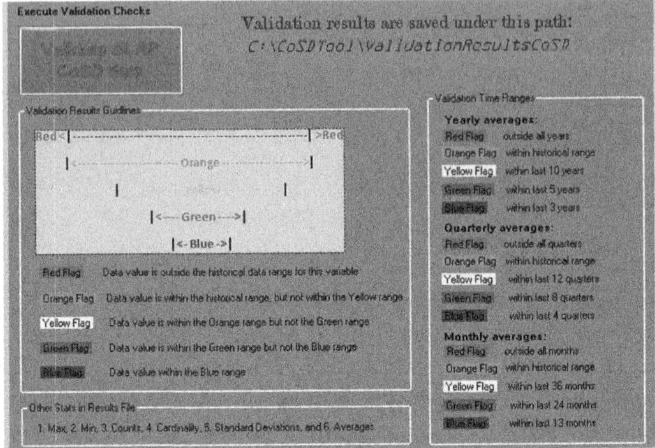

FIGURE 12.1 Flag system representation. *Red flag*, gray in print versions; *orange flag*, light gray in print versions; *yellow flag*, white in print versions; *green flag*, dark gray in print versions; *blue flag*, black in print versions.

rules are analyst-defined rules that indicate when a data point needs attention. Various colors are assigned, for example, a green flag indicates that a data point exists within the expected range (based on the data from the last 3 years). Fig. 12.1 shows a pictorial representation of the flag system. Red flags are for data points that are very far from the average of that data variable. Orange is closer, yellow flags are closer, and so on. Green and blue flags represent the data that have the highest validity. Multiple validation functions were coded that cover certain issues in data values, such as: (1) calculate data values' standard deviation distance from a certain average; (2) check for negative numbers; (3) check for decimal points, zeros, and nulls; (4) allow the analysts to use the federal tool for data validation; (5) perform monthly consistency and redundancy checks.

The technical validation stored procedure (the second part of the engine) validates lookup tables and source tables in the data warehouse. It does three types of validations: identifying duplicates, generating total number of grouped records, and identifying the presence/absence of certain records in a table. Summary statistics (the third part of the validation engine) generates a consolidated statistical report for various tables in the database. The statistical report includes minimum values, maximum values, means, medians, modes, range cardinalities, and standard deviations. Each validation piece is developed using an SQL technique called **CTE (Common Table Expression)**. CTE generates a number of temporary standard results that are all eventually shared as output to the user. Fig. 12.2 shows a code snippet indicating the flag engine working process for yearly green flag data, as an example. The engine's code checks whether the current year's "*value*" data point falls within the range of the last 3 years. If so, then that data point is inserted in the

```
;with CTE CurrentYear Greenflag AS
(
'SELECT ID,VALUE,UNIT,Description
 FROM [CoSD].[DataValues]
 WHERE YEAR IN ( YEAR(GETDATE()))'
)

,CTE Green flag as
(
'SELECT ID,
 Minimumvalue = MIN(VALUE),
 Maximumvalue = MAX(VALUE),
 FROM [CoSD].DataValues]
 WHERE YEAR BETWEEN YEAR(GETDATE()) - 3 AND YEAR(GETDATE()) - 1 '
 )


 INSERT INTO #TEMP1
 'SELECT VALUE,Unit,GREEN as FlagType
  FROM CTE Green flag LEFT JOIN CTE CurrentYear Greenflag
  ON CTE Green flag.ID = CTE CurrentYear Greenflag.ID
  WHERE CTE CurrentYear Greenflag.VALUE
  BETWEEN CTE Green flag.Minimumvalue AND CTE Green flag.Maximumvalue'
```

FIGURE 12.2 Claiming a green flag through common table expression.

```
,cteMedian as ('Get the median for the series of numbers')
,cteMode   as ('Get the mode for the series of numbers')
Select A.GrpByID AS 'CommodityID'
    ,A.LongDescription AS 'Commodity Description'
    ,A.Sourcevalue AS 'Source Description'
    ,RecordCount   = Count(*)
    ,StartYear     = MIN(A.datavaluesyear)
    ,EndYear       = MAX(A.datavaluesyear)
    ,Minimum       = Min(A.VALUE)
    ,Maximum       = Max(A.VALUE)
    ,Mean          = Avg(A.VALUE)
    ,Median        = Max(B.VALUE)
    ,Mode          = Max(C.VALUE)
    ,StdDev        = STDEV(A.VALUE)
    From cteSeriesNumbers A
    LEFT Join cteMedian B on A.GrpByID=B.GrpByID
    LEFT Join cteMode   C on A.GrpByID=C.GrpByID
    Group By A.GrpByID,A.LongDescription,A.Sourcevalue
```

FIGURE 12.3 Statistical functions with SQL.

temporary table "***#GreenFlagTemp***," before being presented to the federal analyst as a green flag.

For any given series of numbers, the **Median and Standard Deviation** method (Evans et al., 2015) is used for finding statistical *outliers*. First the engine identifies the median, then it finds the largest deviated number from the median by finding the difference between the median and each number in the series. The number that is the furthest is identified as *the outlier* (which might be removed from the data set by the analyst). See Fig. 12.3 for an example of statistical functions in SQL. The technical validation engine uses simple SQL

```
SELECT count(*) as [Number Of Duplicate Values]
      ,[Macro_Desc]
      ,[Macro_LongDesc]
      ,[Macro_Value]
      ,[Macro_Source_ID]
      ,[Macro_TimeDimension_ID]
      ,[Macro_GeographyDimension_ID]
      ,[Macro_Unit_ID]
  FROM [CoSD].[Macro_LU]
  GROUP BY
       [Macro_Desc]
      ,[Macro_LongDesc]
      ,[Macro_Value]
      ,[Macro_Source_ID]
      ,[Macro_TimeDimension_ID]
      ,[Macro_GeographyDimension_ID]
      ,[Macro_Unit_ID]
      having count(*) > 1
```

FIGURE 12.4 Finding duplicates example in a data warehouse.

procedures to find duplicates, errors, mismatches, and anomalies (Fig. 12.4 shows duplicates detection code).

Both engines (math and validation) are executed through the federal tool; that is presented next.

5. THE INTELLIGENT FEDERAL DATA MANAGEMENT TOOL

This section provides a comprehensive description of the various services provided by the **federal tool**. The federal data tool is an application designed for analysts and economists at federal agencies for managing, validating, calculating, and streaming data.

5.1 Federal Tool Requirements

Software and hardware requirements were collected from federal teams and management to develop the federal data tool. The main user requirements for the tool are:

1. Analysts should be able to stream data from different sources using the tool.
2. Analysts should have the ability to manage the life cycle of the data in all databases.
3. Important actions taken by users should be logged into the system. If the user deletes some data or updates a value, then such actions need to be tracked in an *action log*.
4. Analysts should have the ability to view data series with different filters.

5. Analysts should have the ability to perform statistical analysis. Details such as data means, modes, medians, and various data occurrences should be available.
6. Analysts should be able to request or add a new data series. They should have means to send an email to data coordinators at other agencies requesting new variables.
7. Ability to import data from excel sheets to the database. Most of the organizational data are present in excel sheets and all of it needs to be in the database in a standardized format.
8. Analysts should have the ability to perform automated validations on data (and use the validation engine).
9. Analysts should be able to update all the agricultural commodity-specific metadata.
10. Based on requirements analysts should be able to change the privacy level of data. Data privacy levels could be public, private, or confidential.
11. A feature to manage all the data migration rules. A **knowledge base** that includes mappings between two sources should be easily manageable. Analysts should be able to add, update, and delete the data migration rules.

The federal data tool was developed, and it currently makes the lives of the analysts easier by managing, validating, streaming, and calculating federal data intelligently. The details of implementation are presented next.

5.2 Federal Tool Implementation

The federal tool has multiple functionalities; all the requirements mentioned are built into functions and the user interface. Figs. 12.5–12.8 show screenshots from the federal tool. Fig. 12.5, for example, shows how federal users can stream the data directly from different sources into their own database. The tool provides a feature to view an existing data series in a database. Fig. 12.6 shows how the tool can perform statistical analysis on the data selected by the user. Users can perform various checks to validate data. User actions, such as viewing, updating, adding, deleting, and resetting data, are common across all the functional areas of the tool (refer to Fig. 12.7). Furthermore, the web-based tool shown in Fig. 12.8 is another powerful data analysis and visualization tool. This tool uses all the data and helps the federal agency analysts and researchers analyze data more efficiently and in a visual manner. The tools allow analysts to export the data to other software packages such as **R**, **Excel**, and **Tableau** (MSDN, 2017).

The tool is currently deployed at USDA agencies, and is continuously under improvement and maintenance. For more details on the code or the availability of the tool, please contact the authors.

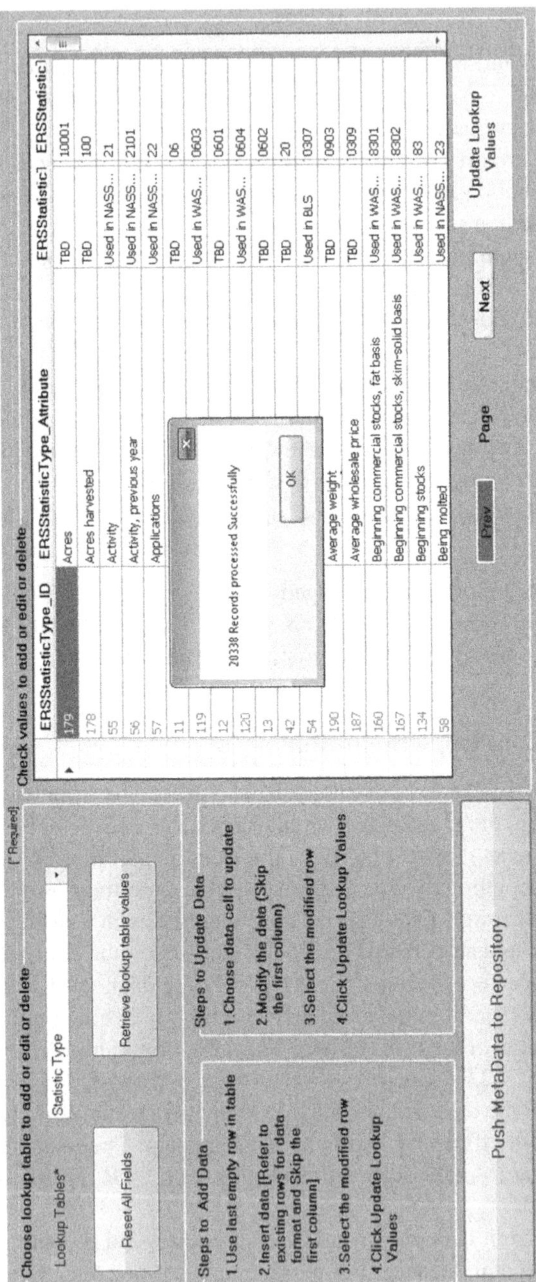

FIGURE 12.5 Managing data streaming into the federal agricultural data warehouse.

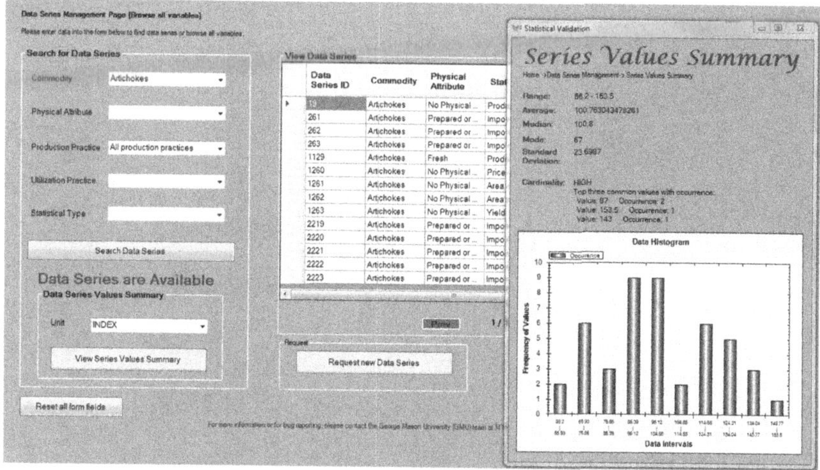

FIGURE 12.6 Statistics and data manipulations through the federal data tool.

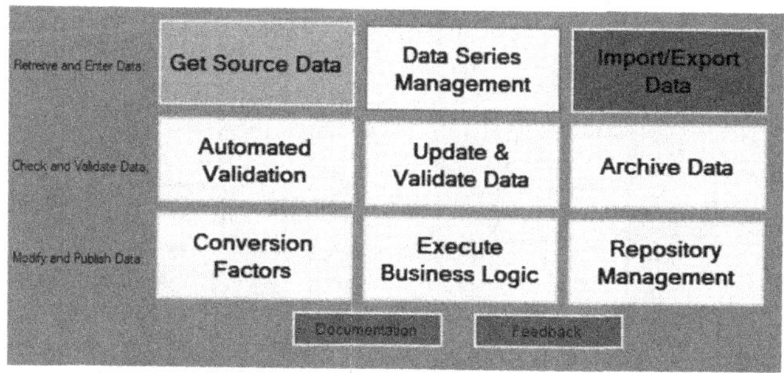

FIGURE 12.7 The majority of tool functions (from the home page of the federal tool).

6. INSIGHTS, EXPERIMENTAL WORK, AND CONCLUSIONS

This section introduces the main keys to addressing federal technical challenges; a list of suggested best practices is presented in detail. In addition, this section introduces the agricultural economists' feedback regarding the proposed engines and tool, as well as other experimental results and insights.

6.1 Experimental Setup and Results

The engines and the federal tool were deployed and used by the federal analysts and the agricultural researchers. Feedback was collected through a survey, shown in Fig. 12.9. All the users were asked to highlight their answers; 20 users were surveyed and the results were collected; the results are presented in Fig. 12.10.

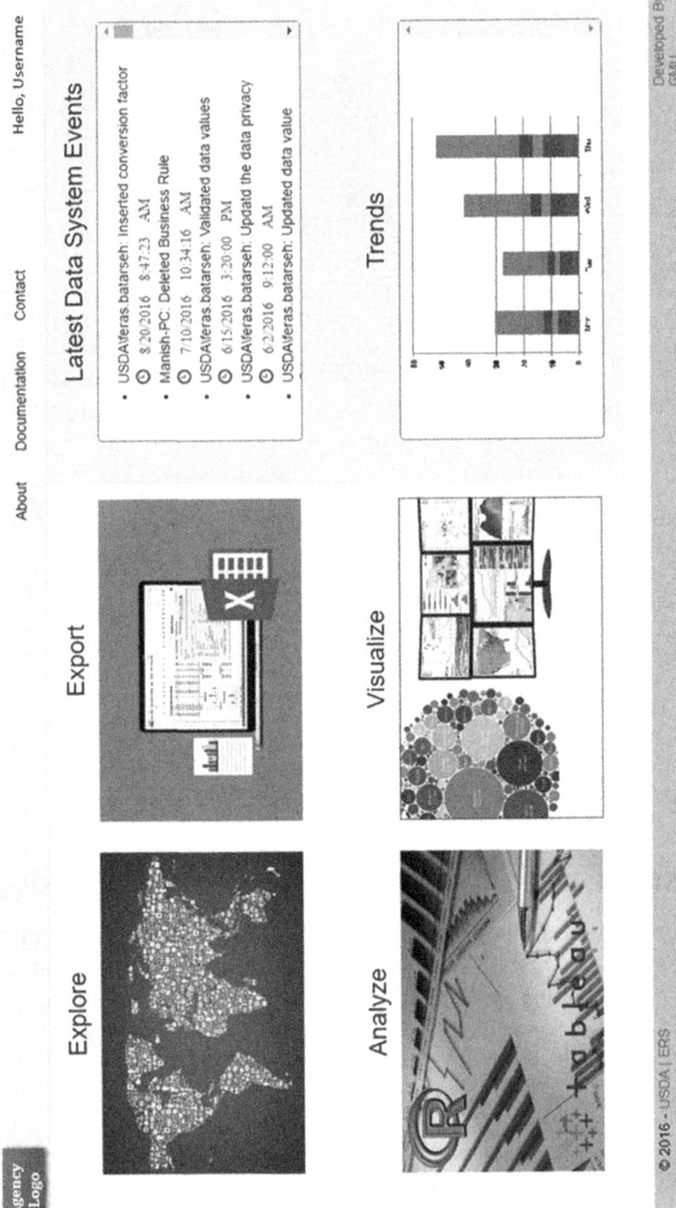

FIGURE 12.8 Visualizing and analyzing data though the web-based tool.

	Please highlight the appropriate score.				
	Very Bad	*Bad*	*Ok*	*Good*	*Very Good*
Rate the usability of the tool:	1	2	**3**	4	5
Rate how the tool matches your tasks:	1	2	3	4	**5**
Rate the easiness of the tool:	1	2	**3**	4	5
Will this tool make your tasks easier (save time and effort):	1	2	3	4	**5**
What do you think of the business rules execution:	1	2	3	4	**5**
Rate the data validation processes within the tool:	1	2	3	**4**	5
Rate the speed/performance of the tool:	1	2	3	**4**	5
Rate error handling in the tool:	1	2	3	**4**	5
Rate your overall acceptance (satisfaction) of the tool:	1	2	**3**	4	5

FIGURE 12.9 Federal tool survey questions.

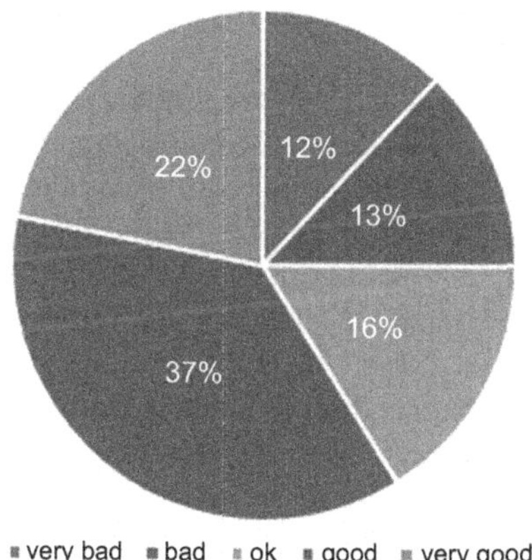

FIGURE 12.10 Federal analysts feedback results.

As an experimental conclusion, a total of **57%** of the analysts provided very positive feedback, which is considered a good starting point for the success of injecting data science into a federal agency. If the practices and technical details of this chapter are followed at any federal agency, it would greatly facilitate the process of injecting an automated data system, which allows for further analysis and advanced algorithms. For more details on the engines, the tool, or the code, please contact the authors.

6.2 Lessons Learnt and Keys to Technical Federal Success

Best practices are listed and discussed in this section; the ones deemed the most important are the following.

User Access Roles: The federal data tool deploys authorization mechanisms for user's access rules (which enhances federal software and data security). User information and access information are maintained in a database,

FIGURE 12.11 Example of automated user access roles.

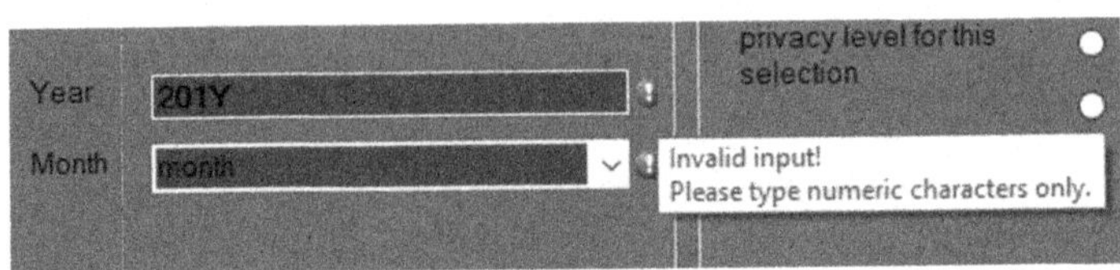

FIGURE 12.12 Regular expressions verifications.

which can be controlled by a "master" copy of the tool. Users can only access or modify data that they are authorized for. For example, "User 14" from the *animal products prices* group is authorized to view only *dairy prices* data. That user will have no access to any other type of data, for example, they would not be able to view *vegetables prices*. A *group–role* table is maintained through the tool and it contains users' personal information and their access rules (refer to Fig. 12.11).

Another federal data science best practice is **Data Entry Validation**. Input data validation techniques have been implemented in the tool; the goal is to restrict nontechnical federal users from inserting invalid data into the system. Power data validations such as *regular expressions libraries (REL)* have been used in the development of the tool. RELs are data patterns that are matched against an input text. Microsoft's .Net framework (MSDN, 2017) provides a regular expression engine that allows such matchings. Federal users get validity messages when they enter invalid data (see Fig. 12.12).

Another very important data science best practice is **Multithreading**. The federal data tool has many important features that are time consuming and could be heavy processes. For example, the validation engine has to consume large amounts of data. Such processes can affect performance if not executed in a multithreaded manner. A thread is defined as an execution path of a program. In the engines, all the complicated and time-consuming operations run on a separate thread, which provides great performance improvements, reduces response time, and increases user satisfaction (Batarseh et al., 2017). The **engines** use the ***Background worker***, a library provided by the .NET framework to achieve multithreading (MSDN, 2017). For example, user interface threads are separated from back-end threads (ensures that the user interface does not freeze). Fig. 12.13 illustrates the execution of a time-consuming process and how that is handled in the interface by letting the federal analyst *run it in the background* and continue working on the tool.

All the engines and the tool have a high level of **Application Integration** as well (with multiple software applications, such as Microsoft Excel,

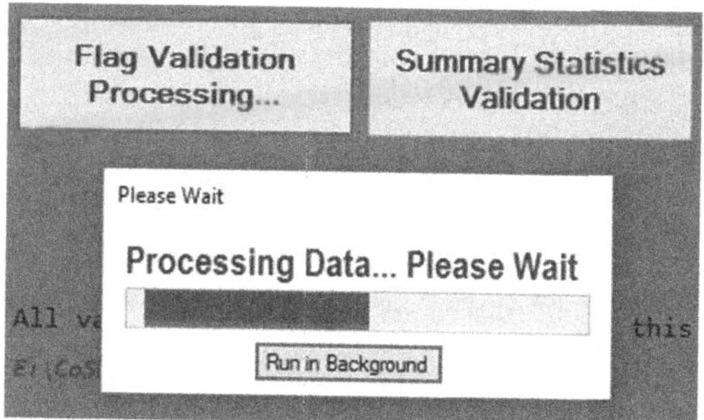

FIGURE 12.13 Multithreading through the intelligent user interface.

Outlook, various Internet Browsers, SQL, Tableau, R, and C#). For example, the federal data tool integrates well with Microsoft Outlook packages to send emails—the ***Interop*** library of the .NET framework had been used for this purpose (MSDN, 2017). The engines also push results to Excel, as previously mentioned, and the web tool connects with Tableau and R (among many other examples).

The engines and the federal data tool handle all the errors by the system or by the user by using exception handling mechanisms of the .NET framework, SQL, and C#. A *try-catch* block is used to separate the code that might be affected by an exception (error). The associated *catch* block is used to handle any unexpected behavior. Analysts interact with the tool as a black box and they do not need to understand the underlying technologies whatsoever. In addition, all the actions taken on the system are recorded in the ***User Action Logs***. The federal data tool logs all the important user actions in the database. Logging of user actions provides significant data maintainability, recovery of mistakes, and overall agency accountability. Any deleted records, for example, are saved in the log, so that it can be retrieved whenever required. This is a very important aspect for federal agencies; it allows for *reversal of actions* and *tracing of federal decisions*.

The engines and the federal data tool help analysts with research and other daily tasks in an organized and professional manner. They also provide access to powerful engines that help in data calculations, analysis, and validation. That ultimately helps analysts take decisions, track actions, forecast, build reports, and influence policy making in agricultural sciences and other areas.

REFERENCES

Anavi-Chaput, V., Arrell, K., Baisden, J., Corrihons, R., Fallon, D., Siegmund, L., Sokolof, N., 2000. Planning for a Migration of PeopleSoft 7.5 from Oracle/UNIX to DB2 for OS/390. A report by Oracle, pp. 147–148.

Batarseh, F., Yang, R., Deng, L., 2017. A comprehensive model for management and validation of federal big data analytical systems. Springer's Journal of Big Data Analytics. http://dx.doi.org/10.1186/s41044-016-0017.

Bertota, J., Gorham, U., Jaegera, S., Choiba, H., 2014. Big data, open government and e-government: issues, policies and recommendations. Information Polity 19, 5–16. http://dx.doi.org/10.3233/IP-140328. IOS Press.

Dynamic SQL, 2016. Available from: https://docs.oracle.com/.

Evans, K., Love, T., Thurston, S., 2015. Outlier identification in model based cluster analysis. US Department of Health and Human Services, pp. 63–84.

Fine, L., Keogh, B., Cretin, S., Orlando, M., Gould, M., 2003. How to evaluate and improve the quality and credibility of an outcomes database: validation and feedback study on the UK cardiac surgery experience. BMJ 326 (7379), 25–28.

Flamos, A., Doukas, H., Psarras, J., 2010. Data validation platform for the sophisticated monitoring and communication of the energy technology sector. The Journal of Renewable Energy 35 (5), 931–935.

Haller, K., Matthes, F., Schulz, C., 2011. Testing & quality assurance in data migration projects. In: Proceedings of the 27th IEEE International Conference on Software Maintenance (ICSM).

Joseph, R., Johnson, N., 2013. Leveraging Big Data: Big Data and Transformational Government. IEEE Computer Society.

Klazema, A., 2016. Data Migration Strategies for Businesses. A Blog. Available from: https://blog.udemy.com/data-migration-strategies/.

MSDN, 2017. C#, Excel, R and other Integration Libraries. Available from: https://msdn.microsoft.com/.

Oracle Corporation, 2016. Move to Oracle Database with Oracle SQL Developer Migrations – Migrating Oracle Databases. A white paper.

Scheier, R., 2015. Data Migration Strategies and Best Practices. TechTarget.

Spivak, D., 2012. Functional data migration. Journal of Information Computing 217, 31–51.

Thalheim, B., Wang, Q., 2013. Data migration: a theoretical perspective. The Journal of Data Knowledge Engineering 87, 260–278.

United States Department of Agriculture's website, 2016. Available from: www.usda.gov.

Woodall, P., Borek, A., Parlikad, A., 2013. Data quality assessment: The Hybrid Approach. The Journal of Information Management 50 (7), 369–382.

FURTHER READING

Batarseh, F., Latif, E., 2016. Assessing the quality of service using big data analytics: with application to healthcare. Journal of Big Data Research 4, 13–24, Elsevier.

Burry, C., Mancusi, D., 2004. How to plan for data migration. In: Proceedings of the 21st International Conference on Advanced Information Systems Engineering, USA.

Chapter 13

Transforming Governmental Data Science Teams in the Future

Jay Gendron[1], Steve Mortimer[2], Tammy Crane[3], Candace Eshelman-Haynes[4]

[1]Booz Allen Hamilton, Washington, DC, United States; [2]Dominion Enterprises, Norfolk, VA, United States; [3]U.S. Department of Defense, Norfolk, VA, United States; [4]NATO Allied Command for Transformation, Norfolk, VA, United States

You don't think your way into a new kind of living. You live your way into a new kind of thinking.

Henri Nouwen

1. INTRODUCTION

Previous chapters of this book highlighted the challenges faced by data science teams. At this point, readers have experienced a healthy dose of information regarding the state of federal organizations' activities in data science. They learned about emerging technological solutions and insights of what types of problems they help solve.

As with most initiatives, leadership will be critical to successful data science projects within the federal government. Leaders can reasonably expect to find and assemble team members who know *how* to do data science. The leaders themselves must shape talent within the team by providing the *what* and the *when*. The diversity of data science is manifest in the variety of skills and experiences today's data science professionals bring to the team. Leading these teams may be a new experience for even seasoned leaders.

Fortunately, one can approach this from two perspectives, situation and personality. The former provides leaders a model to consider tasks and strategy in light of the skills and the commitment within the team at an individual level. Personality provides the "X factor" for team success, an important but hard-to-quantify aspect of individuals. A framework of archetypes can help to aggregate a team of individuals into a smaller collection of personalities.

Federal Data Science. http://dx.doi.org/10.1016/B978-0-12-812443-7.00013-2

Another trait of successful data science teams is continued transformation (Booz Allen Hamilton, 2015). For instance, suppose a team has done a great job creating a data environment for analytic activities. Now that same team must be agile enough to continue its growth to the next phase. Best practices provide a way to discern transformative processes. They aid leaders by providing wisdom during execution and documentation to make the achievements attained from data science momentum more permanent to "retain the gain." These gains may include programmed analytic scripts for data extraction and transformation, data repositories to enable reproducible research, or documented analytic workflows. There are innumerable achievements a data science team will make, and it is important to document them for continuity. Before one may expect to attain success, there is a need to establish leadership within the team.

2. SITUATIONAL LEADERSHIP

Leadership is a timeless topic. Regardless of the time or place, it is essential to organizational success. History provides many an example of good leadership saving the day, and poor or absent leadership leaving situations in shambles. Consider these four generalized statements about leadership:

- Leadership brings about the birth of great initiatives.
- Leadership steadies organizations in distress.
- Leadership propels teams toward goals.
- Leadership is easy to take for granted.

In an informal poll, the authors asked some respondents if they had heard of or used any of these four statements. The last statement garnered the least resonance with respondents. This was not surprising because leadership can be a nebulous topic—one hard to define. Over the years, you perhaps have encountered discussions in the business and academic literature, such as "leadership, nature or nurture"; "leadership versus management"; or "leading emotions, not just people." As a leader in a data science organization, your challenges are no different from those who have gone before you. For that reason, you can learn from history and perspective. Although the topic of leadership for data science could command an entire book, this chapter will present merely one model, situational leadership.

Situational leadership is a model first developed by Hershey and Blanchard (1969). Blanchard later refined his early work into the Situational Leadership II Model (Blanchard et al., 1985). Fundamentally, situational leadership is applying one of four leadership styles based on a person's level of commitment to the task and competence in the skill. Situational leadership proposes that leaders adjust their style according to the situation. Consider this question one may hear within organizations expecting the arrival of a new leader, "What is his or her leadership style?" The answer to that question is not a single, fill-in-the-blank

TABLE 13.1 Applying Situational Leadership Styles According to Commitment and Competence

Style	Commitment Level	Competence
Directing	High	Low
Coaching	Low	Low
Supporting	Low (but growing)	High
Delegating	High	High

response; rather, it depends. Table 13.1 shows the four leadership styles, as well as their associations to commitment and competence, as presented in the models (Hershey and Blanchard, 1969; Blanchard et al., 1985).

Leaders apply one of the four leadership styles listed in the table after observing the commitment and competence levels associated with the person in the situation they observe. Here are some examples of situations along with situational leadership styles:

- **Directing** works well in crisis as well as with someone new to a task who has earnest commitment to the new situation but lacks skill. This style helps the team member get on with the work because the leader provides direction on tasks.
- **Coaching** helps a team member who is growing in skill but may be disillusioned with the task having become familiar with it, and perhaps losing confidence. Consider the football coach who encourages his athlete to run one more lap before stopping for the day. Coaching is useful here because it blends training, direction, and mentoring.
- **Supporting** is appropriate when someone has grown in skill to take on a task himself or herself but may still be developing a sense of how that work applies to the greater team. The leader is present through support of resources or time.
- **Delegating** is a style best used when their team member truly "has the stick." The leader can articulate the task requirements and the skilled, committed person will take care of it.

At this point, realize that different people have not only different skills, but also different skills in relation to certain types of tasks. Someone may be an expert in one task, whereas that same person may be required to perform another task in which he or she lacks the confidence and skill to achieve full success. Situational leadership provides a way to size up the task, the person, and the situation and respond with an appropriate leadership style.

Situational leadership applies particularly well to the data sciences. Data science is a burgeoning field. Organizations, people, and leaders are still "finding their way." It is a mix of mathematicians, scientists, and computer scientists.

This diverse collection of people calls for situational leadership in building, creating, and leading a data science team. In some industries, people follow a fairly well defined path of growth. They may have evolved in a nearly predictable way to their current positions. Data science is different. One observes people arriving from many different backgrounds.

Consider the plight of a data engineer who is an expert in data warehousing but now is on a data science team with roles that go beyond merely cataloging and retrieving data. This engineer will need to work with data in real time, which can be very new to people steeped in older data engineering methods. The data science field also finds computer scientists who are skilled in programming and are now learning the challenges of modeling. The mathematical and statistical realities of model building are new to them, regardless of the number of hours they have writing C++, Python, or Java code. Drew Conway (2010) captures these types of realities very well in his popularly referenced Venn diagram for data science. It supports the idea of situational leadership based on a person's competence and confidence. Data science finds mathematicians, scientists, and statisticians who understand the uncertainties in modeling a phenomenon. However, many of them will find themselves working in a new environment at larger scales or faster speeds than they ever have before.

As a leader of a data science organization in the federal government, expect your team to come from diverse backgrounds with broad skills and experiences. Readers can imagine the variety of experiential bases their team members will have. These may include program management, developmental and test engineering, financial or contract analysis, data warehousing, information technologies, and other areas. The leader's job is to provide the catalyst and the environment to transform these classic government backgrounds into a data science capability. This includes growing in them the key skills of data science such as computer science and machine learning, statistics, foundational mathematics such as linear algebra, as well business domain awareness and communication skills (Viswanathan, 2016). Within the Department of Defense, the Naval Postgraduate School (2017) has updated its successful curriculum to include a focus on data analysis and advanced data analysis using contemporary computational tools such as those found in industry. Nurturing this transformation in people requires leadership—at an individualized level—to help galvanize them as a team. One may sum up the leadership requirements of data science with this mantra:

> *Lead people by assessing their individual skills for the particular task according to the level of engagement that they show for each situation.*

In addition to leadership based on competencies, a good leader can consider a team's personalities as well. The archetypes framework presented in the next section provides one such means for doing that.

3. ARCHETYPES

This section introduces four data science archetypes, which help account for individual and team personality while establishing leadership. This is important because personality provides the "X factor" of the team. This "X factor" is a component that includes many intangibles that are nonetheless important for successful teams. Leadership is not clean and simple. It does deal with people after all. Given the unique nature of each person, archetypes provide leaders a framework to address the intangibles by categorizing roles and functions into a few concrete personas.

The four archetypes presented in the following list demonstrate the core (minimal) set of functionality that data science organizations expand over time. The organizational structure of these archetypes should be relatively flat and autonomous to facilitate idea generation and open collaboration. In the basic framework, consider these four archetypes and their data science functions:

- The **strategist** considers how the deliverable will provide value for stakeholders and determines data product features, how to market the benefits, and expected users.
- The **designer** creates algorithms, models, and visualizations that balance effectiveness, accuracy, and simplicity.
- The **tester** uses a keen eye to identify the flaws of a product and present recommendations for improvement.
- The **engineer** structures systems and data flow so that the designer and tester have access to systems and data.

3.1 Mapping Archetypes to Prior Career Paths

Archetypes describe roles of individuals, but they do not describe who exactly should fill those roles. Given that data science is a relatively new and growing field, individuals on a data science team have most likely held other titles in past careers. In this section, we will map out some job positions found in governmental settings that correlate with the four data science archetypes.

Project managers are well suited to be the strategist for a data science team. They have the experience of working with subject matter experts and gaining knowledge that heightens the nuances of a more data-centric product. A less commonly thought of prior career path for the strategist would be a technical writer or implementation consultant (information technology (IT) administrators within government settings). These individuals often bring unique insight on user behavior and the pitfalls to achieving more optimal adoption, which are mission critical skills for the strategist. Regardless, leadership is essential to the strategist archetype to provide a strong vision and purpose for the team (Vries, 2013).

The designer will probably have a background in statistics, economics, or computer science. However, Patil (2011) notes that great designers of data products can come from a variety of roles. His team has had great breakthroughs from chemists, oceanographers, and even neurosurgeons. A thorough technical interview can accurately gauge the potential of a candidate for the designer role.

The tester's main contribution is quality. A number of disciplines possess the key skills to fill this role. Test and evaluation engineers, quality assurance engineers, and computational financial analysts are but a few of the domains that can translate their skills into the more specialized field of data science. Alternatively, someone who has functioned in the designer archetype can serve in the tester role. This could be a person actively working as a designer on another data science project, or someone who has transitioned out of that capacity and into roles as subject matter expert or reviewer of others work.

The engineer is responsible for IT systems and data pipelines. Those individuals best fitting this role might have experience as a database administrator or a systems administrator. Another career path might be back end developer. Back end developers are responsible for the configuration of database, server, and application code to support web applications, which means they have skills required of an individual filling the engineer role.

3.2 Archetypes in the Federal Government: An Example

Of the three topics presented in this chapter, the idea of an archetype is likely the most foreign to a reader from a governmental agency. You have probably encountered the four essential archetypes as well as their functions within the government. This example provides a concrete way to consider some characteristics you are familiar with in a data-centric context. An organization within the Department of Defense exhibits the elements of each archetype. It may or may not have been a conscious decision, but the archetypes are present.

This data science team is responsible for analysis that helps senior leaders make fiscal and policy decisions. The team works at the strategic and operational levels. A natural evolution has occurred where analysis formerly created and stored in Microsoft Excel is now becoming manifest in analytic software products the team builds. The natural aspect of the evolution is the increased use of cloud computing and web-based products within the department. The team has a strategist by the way of a government project manager who oversees the deliverable. This person has placed an emphasis on the value proposition of the resulting analysis. This focus led to a web-based analytic tool.

Two designers are also on the team. They are model builders, having formal educations in the mathematical domain. One of the designers plays a dual role by helping the strategist to bring his experience of software development into the strategy by helping frame the discussions as if the tool were going to compete in a marketplace. In a way it is. Constrained resources result in many of the same market forces from industry that drive the analytic product development. Fortunately, for the team, the other designer serves as a tester and is very good in rigorously conducting verification and validation of the models. In fact, they assist another analyst who is serving more or less full time as a tester. The tester continually takes new, proposed releases and tests them from a user perspective for flaws.

The tester also works closely with a small group of people on the team who are responsible for the data flow. One in particular operates in a way that readily enables examination of various types of data structures and frameworks. In total, these six people, who are part of a larger team of a dozen or so people, represent the functionality described in the archetypes. Having looked at an actual example, you can now consider how the archetypes allow leaders to manage risk and growth.

3.3 Archetypes, Risk Mitigation, and Growth

Employing these archetypes provides a framework for both risk mitigation and growth. Risk mitigation comes from building redundancy into the team structure. For example, if the engineer were unavailable, then the designer could perform day-to-day engineering duties. Designers will have intimate knowledge of the data pipeline because they are a consumer of it. Each archetype works closely with other archetypes to enable knowledge transfer and maintain day-to-day operations. A fully staffed team would ideally have at least one of each archetype; however, in resource-constrained environments, one person can serve in multiple archetype roles. Contractors and technology vendors are an excellent use of resources that can provide new teams experienced individuals to arbitrate the ideas and concerns when creating architectures and processes.

In addition to mitigating risk, it is important to consider stretch roles and growth opportunities for team members, for example, encouraging engineers to take stretch roles to learn complex elements of model building from the designer. It is also encouraged that testers take stretch roles to learn from strategists on setting a vision, defining the mission, and enabling a strategy based on the data product. Personal growth and development is key to the team's success (Booz Allen Hamilton, 2015).

The previous paragraphs have emphasized redundancy and team development through shared stretch roles. Now consider the opposite of team contraction, that of team expansion. These same archetypes scale well because of the scope of their responsibilities. The designer and tester scale well to meet manning requirements of the project. Keep in mind that leaders should scale these two roles in similar proportions so that all designs may be thoroughly tested. The strategist and engineer roles typically do not scale as quickly. In some cases, features may require more labor hours but do not materially change strategy or require reengineering the data pipeline. Use discretion when expanding strategy and engineering roles. These changes could mean that a new project is being undertaken, which warrants a completely new data science team. Similar to other teams, data science team output diminishes as the team grows larger because of social loafing and loss of coordination.

4. BEST PRACTICES

Earlier chapters of this book present numerous examples and vignettes of data science practice among departments and agencies of the federal government. Leadership and transformation lie behind the successes in meeting challenges and furthermore retaining them in a data science culture. Having read the vignettes, what can you take away from those accounts? You can consider them thoughtfully, pluck ideas most suitable to your situation, and establish the culture within your own data science team. In essence, you can establish best practices by learning from others.

4.1 Creation of a Best Practice

A *best practice* is a technique or methodology that emerges from a proven, rigorous process of peer review and evaluation resulting in effective outcomes relative to some subject, process, product, or population. Often, the development of best practices takes place over many years. Data sciences in the federal sector will be no different. Fortunately, leaders can rely on some guiding principles to develop and refine best practices for federal data science:

- Best practices are repeatable and sustainable, producing consistent and tangible results across a variety of environments.
- Best practices clearly link effects to the practice under evaluation.

Consider a commonplace best practice culture that has emerged from the automobile industry and how it pervaded into our popular culture. Generally, automobile owners agree that getting the best performance from their vehicle requires regular maintenance: belt and hose replacements, battery and wiper blade checks, tire rotations, and oil changes. Auto maintenance is a common example of a best practice. Failing to perform these best practices will eventually cause the vehicle to fall into disrepair. The question to consider is, "How did these best practices solidify themselves as 'truths' within the culture?" They were not set forth as policy and they did not emerge instantaneously.

A best practice is most effective when one considers not only results but also sustainability and continuous refinement of the approach relative to stakeholder interest. An organization that establishes a best practice culture routinely questions *what* the enterprise is doing and *why*. This introspective approach occurs throughout the project and at all levels of the organization. It can transform gains achieved by your data science teams into sustained organizational success.

The concept of a best practice for data science could take up an entire volume, but space allows for the presentation of a particularly important topic for data science best practice—collaboration.

4.2 Intra- and Interagency Collaboration

According to a GAO study (US Government Accountability Office, 2012), federal agencies use a variety of mechanisms to implement interagency collaborative efforts (which are also applicable to intraagency collaboration). These implementations include shared databases and web-based portals; liaison positions where an employee from one organization works primarily with another agency; and shared leadership where multiple agencies agree to accountability for an initiative to benefit from data collection, analysis, and lessons learned.

These practices are worthy of consideration in data science. The communication and coordination challenges surrounding data must not be underestimated. Numerous questions and concerns emerge in organizations regarding data access, usage, and reporting. That 2012 study identified areas where opportunities existed for federal agencies to transform their processes by collaborating across and within agencies. The data show benefits from the following key practices when implementing collaborative mechanisms:

- **Outcomes and Accountability**
  Have short-term and long-term outcomes been clearly defined? Is there a way to track and monitor their progress?
- **Bridging Organizational Cultures**
  What are the missions and organizational cultures of the participating agencies? Have agencies agreed on common terminology and definitions?
- **Leadership**
  How will leadership be sustained over the long term? If leadership is shared, have roles and responsibilities been clearly identified and agreed upon?
- **Clarity of Roles and Responsibilities**
  Have participating agencies clarified roles and responsibilities?
- **Participants**
  Have all relevant participants been included? Do they have the ability to commit resources for their agency?
- **Resources**
  How will the collaborative mechanism be funded and staffed? Have online collaboration tools been developed?
- **Written Guidance and Agreements**
  If appropriate, have participating agencies documented their agreement regarding how they will be collaborating? Have they developed ways to continually update and monitor these agreements?

Collaboration is important because leaders and team members have limited bandwidth. This makes your team a group of specialists in the problems they have solved and the algorithms they have successfully implemented. These same algorithms are reusable, so leverage the work of others through collaboration. D.J. Patil addressed the 2015 Strata-Hadoop Conference just days after becoming the United States' first chief data scientist. Patil (2015) said, "Data science is a team sport, and we can't do this without you...We really need your help...You don't have to be a U.S. citizen. You don't have to relocate to D.C. There's all sorts of ways to jump in."

A key idea presented by D.J. Patil is the notion of not reinventing the wheel. So many algorithms are applicable across multiple contexts. A number of collaborative efforts result in today's federal agencies realizing they can adapt the work product from previous efforts and simply tailor it to the needs of the current project. This results in saving time, funding, and other resources. As a lesson learned, this is also a best practice, from which interagency sharing is slowly benefitting.

5. CONCLUSION

As the federal government increases its use and reliance on data science to tackle its problems, the need for effective leadership is essential. Leadership is often the critical difference between success and failure of an initiative. The burgeoning nature of data science will likely pose challenges for even seasoned leaders during execution. This is in part due to the diverse collection of people that will form into data science teams within the government.

Fortunately, models such as situational leadership help in assessing the task and skill mix to recommend a suitable leadership style. The essence of this leadership framework, first presented by Hershey and Blanchard, is to use a leadership style customized to a person's skills as well as their competence with that skill. Use the approach in discrete situations—again, customizing the leadership style used on individuals within the context of the task they are executing and within the environment where the task occurs.

As flexible as situational leadership is, there are also frameworks such as archetypes to simplify an organization's complexity to promote ideation and growth. The simplification in this framework is aggregating the various individuals into one of four broad archetypes essential for a data science team. Consider these archetypes as combinations of personalities that help to generalize the team and consider them in higher-level functions. This chapter has also provided an example inspired by a government team to help demonstrate how aggregating the individuals by archetypes condenses the many aspects of individual personality in a manageable way.

After forming a federal data science team, continued transformation is possible by promoting a culture of best practice. This chapter defines the term best practice and notes it requires a history of proven, reviewed, and evaluated outcomes. Although this takes time, federal data science will most likely use the lessons learned that are available to jump start the process of leveraging and creating best practices. Two guiding principles exist to encourage leaders in this pursuit. Along with best practices, collaboration is a way to consider how to reduce redundancy in creating products and workflows within a data science team.

The future of data science in the federal government is very promising. The modern government is both a large supplier and user of data. It is hard to say with certainty what the future holds in terms of the particulars of data science within the federal space. However, if the commercial sector is any indicator, then leaders of data science teams can expect rapid evolution of technology for use on larger and larger data sets. They can also expect their talented and ambitious team members to strike out and find ways to attack the problems algorithmically and quickly. This requires a leader to promote trust of the data science team—between not only the leader and the team, but also among the team and the organizations they support. The future of data science is interesting to those who work within the domain—it is reasonable to assume that those served by data science may be skeptical at first. It will take time for the capabilities to become commonplace and result in fewer barriers to acceptance.

ACKNOWLEDGMENT

Thank you to Mr. Daniel Killian, a data scientist with the Department of Defense, for serving as a reader of this chapter. Daniel, a graduate of Rensselaer Polytechnic Institute and New Mexico State University, currently works as an Operations Research Analyst for Headquarters, United States Marine Corps, Program Analysis and Evaluation, and has worked for the US Army Training and Doctrine Command Analysis Center.

REFERENCES

Blanchard, K.H., Zigarmi, P., Zigarmi, D., 1985. Leadership and the One-minute Manager: Increasing Effectiveness through Situational Leadership. Morrow, New York.

Booz Allen Hamilton, 2015. Tips for Building a Data Science Capability: Some Hard-Won Best Practices and (Surprising) Lessons Learned. Available from: https://www.boozallen.com/content/dam/boozallen/documents/2015/07/DS-Capability-Handbook.pdf.

Conway, D., September 30, 2010. The Data Science Venn Diagram. ZIA: Blog. Available from: http://drewconway.com/zia/2013/3/26the-data-science-venn-diagram.

Hershey, P., Blanchard, K.H., 1969. Life cycle theory of leadership. Training and Development Journal 23 (5), 26–34.

Naval Postgraduate School, March 12, 2017. Operations Analysis – Curriculum 360. United States Navy. Available from: http://www.nps.edu/Academics/GeneralCatalog/414.htm#o441.

Patil, D.J., 2011. Building Data Science Teams: The Skills, Tools, and Perspectives behind Great Data Science Groups. O'Reilly Media, Sebastopol, CA. Available from: http://www.datascienceassn.org/sites/default/files/Building%20Data%20Science%20Teams.pdf.

Patil, D.J., February 19, 2015. 'Data Science is a Team Sport': D.J. Patil Spends First Day Pitching Silicon Valley on Joining Government. Nextgov: Blog. Available from: http://www.nextgov.com/big-data/2015/02/data-science-team-sport-dj-patil-spends-first-day-pitching-silicon-valley-joining-government/105648/.

United States Government Accountability Office, September 2012. Managing for Results: Key Considerations for Implementing Interagency Collaborative Mechanisms. Government of the United States of America. Available from: http://www.gao.gov/assets/650/648934.pdf.

Viswanathan, S., 2016. Data science: exciting days ahead. Human Capital 20 (2), 34–35.

Vries, M.F.R.K., 2013. The Eight Archetypes of Leadership. Harvard Business Review. Available from: https://hbr.org/2013/12/the-eight-archetypes-of-leadership.

Afterword

For data science at government, the future is now. This book introduces a number of solutions and guidelines for implementing data science, injecting Artificial Intelligence (AI), and deploying advanced analytics at different sectors of the US government. The authors evaluate existing systems at federal, state, and city agencies in America as well as other countries (such as Canada, France, Spain, and China). This book aims to serve as a guide for federal engineering teams, data science consultants, and government contractors in applying analytics and their best practices to existing workflows. For example, through the methods presented in this book, intelligent algorithms could be applied to all phases of data management (i.e., managing agricultural economics data coming from *farms* all the way to consumers' *forks*).

However, to achieve the required data science dynamism at government, agencies need to address the instant need for increased expertise in the fields presented in this book. Agencies should invest in data sooner rather than later (for example, by connecting with American universities for training and development), or they will miss the unstoppable train of data science that most organizations are riding. It is not necessary for all federal analysts to become data science experts. However, it is clear that some persons at different levels of an agency should advocate for it. Most importantly, federal employees should invoke it in the *culture* of their divisions and be *AI champions* at their agencies. The AI solutions presented should not be solely in the hands of specialists and software experts; rather, they should be igniting systematic changes for all employees. Such changes should improve existing manual processes and eventually advance an entire agency.

This book is not pushing for any political view or agenda; all the work presented is based on scientific analysis and experimentation. Authors of this book come from different backgrounds (such as the federal government, international organizations, universities, and the private sector) but agree on one notion: *the necessity of more data science at the US government*. There are multiple books published in the areas of AI, federal systems, agricultural economics, data, and intelligent software; nevertheless, this book is at the intersection of these topics, an area that was not explicitly explored prior.

Feras A. Batarseh and Ruixin Yang
George Mason University
Fairfax, Virginia
2017

Index

'*Note*: Page numbers followed by "f" indicate figures, "t" indicate tables.'

M

N

O

P

R

S

T

U

Printed in the United States
By Bookmasters